GE IN S

CAL ON S

To my father, John Colin Madgwick, who encouraged and nurtured
my enthusiasm for science and research, and my late mother,
Valerie Lorraine Madgwick (nee Jenkins),
a truly inspirational person.

GEOGRAPHICAL INFORMATION SYSTEMS

AN INTRODUCTION

Julie Delaney

OXFORD
UNIVERSITY PRESS

OXFORD
UNIVERSITY PRESS

253 Normanby Road, South Melbourne, Victoria 3205, Australia

Oxford University Press is a department of the University of Oxford.
It furthers the University's objective of excellence in research, scholarship,
and education by publishing worldwide in

Oxford New York

Auckland Bangkok Buenos Aires Cape Town Chennai
Dar es Salaam Delhi Hong Kong Istanbul Karachi Kolkata
Kuala Lumpur Madrid Melbourne Mexico City Mumbai Nairobi
São Paulo Shanghai Singapore Taipei Tokyo Toronto

with an associated company in Berlin

OXFORD is a trade mark of Oxford University Press
in the UK and in certain other countries

National Library of Australia
Cataloguing-in-Publication data:

Delaney, Julie, 1970–.

Geographical information systems: an introduction

ISBN 0 19 550789 4.

1. Geographical information systems. I. Title.

910.285

Edited by Marta Veroni
Indexed by Max McMaster
Text and cover designed by Derrick I Stone Design
Printed in China through The Bookmaker International Ltd

Contents

List of figures

Preface

A note to the reader

Aims of this book

This book introduces the key fundamentals of Geographical Information Systems (GIS) and serves as a precursor to formal GIS training or as a reference within a structured unit or course.

The level of knowledge in this book will allow the reader to progress towards using GIS software by developing an understanding of its principles, concepts, and practices. In essence, this book aims to remove the mechanical approach of feeding data into a computer, typing commands, and achieving a result without understanding what GIS is about.

This introductory text attempts to provide a simplified, jargon-free explanation of GIS. Most books aimed at novice GIS users are produced or supported by GIS software companies. One can thus easily become a software-specific user and lose sight of the much wider GIS field. This book will encourage you to develop a software-independent foundation, enabling you to speak and think as a more versatile GIS user.

To progress from this book to one of the many available software packages you will need the appropriate dictionary in order to translate generic understanding into software-specific terms, and to master commands. Using the generic approach should ensure relative ease in moving between software packages.

The second aim of this book is to enhance the understanding of GIS concepts through the use of examples, thus making GIS relevant and applicable to a multitude of uses. The exercises are based on data gathered from a Western Australian property called Jarrahlea.

A third aim of this book is to provide an insight into the experiences of GIS users. This commentary is the result of a survey of a group of people who have been in the GIS workforce less than five years and includes employees from both the public and private sectors. It is helpful for you, as a new GIS user, to listen to others who have recently made the transition from learning to application. The survey was limited to only 30 respondents and asked them to comment on their GIS experiences, using the chapter headings of this book as a guide to areas of interest.

Interesting facts and figures regarding GIS education in Australia

The survey revealed that almost 17% of respondents had received commercial software training offered by a software company. These courses offer focused training in the company's software package at a substantial cost. A quarter of respondents had a formal university GIS qualification, such as a Diploma. Over 40% of respondents had gained GIS experience as part of a larger field of study in courses at universities or colleges within a Bachelor of Science or Arts course. An interesting finding was that the remaining 17% of respondents were self-taught using a multitude of resources.

Regardless of the method used to acquire GIS skills, only two thirds of respondent GIS users had made a conscious choice to incorporate GIS into their career path. Many seemed to be driven by the needs of their employer, or by requirements to undertake spatial data analysis.

A survey conducted in the late 1980s would have indicated that the majority of GIS users were self-taught, with a background in cartography or the mapping sciences, or were taught to use a specific software package. However, currently the field of GIS interest is growing rapidly and is being incorporated into a variety of learning arenas.

Why this book was written

There are a number of reasons why this book was written, apart from being an aid for self-teaching GIS.

There is a rapidly growing body of GIS literature. Many of these publications are aimed at experts rather than novices, or deal with specific research issues. There is a clear distinction between students who are familiar with the computational concepts and spatial science students. The former student makes the transition to GIS software almost innately, but may not fully understand spatial or geographical concepts. The spatial science students may not be computer experts, but they can gain significant advantages from GIS application in their field of expertise.

> I like geography. It is the only field I want to study. Why should I have to do computer science too?
>
> *Undergraduate student, after a two hour computer*
> *laboratory which took six hours to complete*

This text caters for the person with a limited (or non-existent) computing background. Many current GIS texts are hampered by the use of complicated and confusing jargon.

> Isn't there a simple GIS for dummies book I can read quickly instead of tackling these involved technical manuals? I really need a quick, simple to understand, overview of GIS.
>
> *Frustrated mature aged student studying part-time*

This book attempts to reduce this overwhelming terminology and lets you deal directly with the concepts of GIS and geographical analysis. All GIS-related phrases printed in bold can be found in the glossary at the end of this book.

Unfortunately, many GIS units and courses within Australian universities are operating under diminishing budgets. This means that expenditure on computing hardware, software, tutors, and demonstrators must be rationed. One way in which this trend can be alleviated without sacrificing the impact of applying GIS is to use a combined GIS text and case study workbook.

> I wait for a computer terminal because I don't want to share a computer in the laboratory class. Then I sit down in front of the machine and type commands. I don't know why I use this command here and that command there. I do it because it tells me to in the course notes.
>
> *Student found in the computer laboratory at 2:00 a.m.*

This introductory text allows students to appreciate GIS investigations without the need of computing resources.

Distance learning is a reality. The Internet and other audio and visual aids enable spatially isolated people to undertake studies. These aids can be expensive and are sometimes plagued by technical computing problems.

> I have been given permission to study GIS in my work hours. I live in Karratha. There is no way that the company will pay for me to go to Perth for practical classes. If it is going to be coming from my pocket I want to know that this is the career path I want to take.
>
> *Independent student with a use for GIS but no direct access to facilities*

A book is a relatively cheap alternative and may serve as a means for evaluating the appropriateness of pursuing GIS study at a tertiary institute.

Most GIS texts are related to experience elsewhere in the world. This text is unique as it employs Australian examples. This does not mean that the international body of GIS students will not benefit from using this book, rather that it offers a different context.

> Miles, kilometers, colors . . . the spelling, the examples—everything is American. If there are no good Australian GIS examples then GIS in Australia must be second rate or boring.
>
> *Disillusioned Australian GIS student after completing a*
> *demonstration provided by a software company*

This, of course, is not true. The GIS community in Australia provides a rapidly growing and exciting learning environment.

How to use this book as a teaching aid

This book can be used as a stand-alone text or as a supplement to a practical GIS course. It allows you to develop a working background in the basic concepts of GIS in the absence of a course. You will not, however, be a capable GIS user until you have gained practical experience with GIS software. The reader of this text will hopefully have a broader understanding of GIS than a person learning GIS based on one software package.

Following this introduction with a practical course would be advantageous. Reading the material and completing the exercises prior to sitting at the keyboard will ensure that you have a rudimentary understanding of what the software is doing rather than treating the GIS software as a 'black box' to do analyses. GIS is a tool and always will be a tool. The success and applicability of the end product hinges on the users' knowledge of the principles of GIS, the data and the processes involved.

Do not sit beside the computer with this book. It will not help; the jargon of the software will confound the understanding of the basic principles. Read the book and then relate it to the version of GIS offered by the software package.

Where does this book fit in the range of current GIS literature?

GIS textbooks are plentiful nowadays, due to remarkable growth in usage of GIS since the early 1980s. This book is a basic guide to GIS. Other guides are cited throughout this book (see References and further reading), however, they do not offer the emphasis on exercises or the application examples as found in this book. It is also a workbook, but it differs from other GIS workbooks in that it is not linked to a software package.

This text does not cover specific topics that may be crucial to GIS experts, technicians, researchers, or specialists. References to texts aimed at these more advanced levels are given. These may be of interest after this introductory text has been completed and the user has had hands-on GIS experience.

You are about to embark on the voyage of GIS discovery. With diligent application you will navigate the concepts and steer through the exercises, but don't get becalmed trying to learn GIS based on one software package!

Acknowledgments

The author would like to thank all the survey respondents, especially those who permitted the publication of their comments: Narah Stuart, Matthew Aylward, Piers Higgs, Werner Runge, Nick Middleton, and Nicholas Welch. The author also gratefully acknowledges the contributors to Appendix 2, Australian GIS Stories: David Smith, Toni Furlonge, Tim Molloy, Carl Bennet, Andrew Burke, and Robin Piesse.

Many thanks to John Madgwick and Colin Madgwick for providing invaluable help by proof reading and commenting on drafts of this manuscript, as well as providing support and encouragement. Gratitude is also due to the referees, for their valuable input when the book was in draft format.

Finally, Craig and William Delaney are acknowledged for their moral support.

Abbreviations

(many of these relate to abbreviations used in Appendix 2)

ANZLIC	Australia New Zealand Land Information Council
CAD	Computer Aided Design
CAS	Catchment and Agricultural Services
CSI	Critical Skill Index
DCDB	Digital Cadastral Data Base
DEM	Digital Elevation Model
DMBS	Database Management System
DNRE	Department of Natural Resources and Environment
DOCIT	Department of Computing and Information Technology
DOLA	Department of Land Administration
DTM	Digital Terrain Model
FFDI	Forest Fire Danger Index
GBC	Goulburn Broken Catchment
GDA94	Geocentric Datum of Australia, 1994
GIS	Geographic or Geographical Information Systems
GPS	Global Positioning System
Id	Identification number
IT	Information Technology
km	kilometres
LID	Land Information Directory
LIS	Land Information System
LISAC	Land Information System Advisory Committee
LISAPC	Land Information System Administrative Policy Committee
LISEMC	Land Information System Executive Management Committee
LISSC	Land Information System Support Centre
m	metres
MGA94	Map Grid of Australia, 1994
NFS	Network File System

PC	Personal Computer
SCDB	Spatial Cadastral Data Base
SIG	Special Interest Group
TCP/IP	Transmission Control Protocol over Internet Protocol
3D	three dimensional
TIN	Triangular or Triangulated Irregular Network
TM	Thematic Mapper
2.5D	two and a half dimensional
UTM	Utility Transformation Manager
WA	Western Australia
WAGCPC	Western Australian Government Computing Policy Committee
WALIS	Western Australian Land Information System

CHAPTER ONE:
The Gist of GIS

GIS constitutes an integrated tool box for spatial data input, storage, management, retrieval, manipulation, analysis, modelling, output, and display.

GIS has many components, including hardware, software, data, and personnel. The interactions of these components give GIS a suite of functionalities, which recommend its use over traditional techniques in a wide variety of applications.

What is GIS?

Geographical (or geographic) **information** systems have existed since the 1960s, although the techniques used by GIS predate this. The popularity of GIS and GIS **tools** was initially inhibited by the expense and expertise required to set up a system. In recent years (the 1980s and 1990s) the use of GIS has grown dramatically and this has been attributed to a number of factors, some of which are listed below:

- the decreasing cost of computers, GIS software, and pre-captured data;
- increasingly user-friendly software;
- the realisation of the potential benefits of using GIS;
- the increasing availability of spatial data in digital format;
- the appearance of GIS education and training programs in universities, colleges, and schools, providing a GIS work force;
- the development of complementary technologies, such as remote sensing and global positioning systems; and
- the growing need for conducting spatial decision-making in a more scientific and accountable fashion.

Definitions of GIS range from the very technical to the simplistic, from functional to abstract, detailing the components or describing the elements. Example definitions follow: GIS is 'an institutional entity, reflecting an organizational structure that integrates technology with a **database**, expertise and continuing financial support over time' (Carter 1989), in addition to being 'a powerful set of tools for collecting, storing, retrieving at will, transforming and displaying **spatial data** from the real world for a particular set of purposes' (Burrough 1986). GIS is also 'a database system in which most of the data are spatially indexed, and upon which a set of procedures operated in order to answer queries and spatial entities in the database' (Smith et al. 1987) and 'the organized activity by which people:

- measure aspects of geographic phenomena and processes;
- represent these measurements, usually in the form of a computer database, to emphasize spatial **layers**, entities, and relationships;
- operate upon these representations to produce more measurements and to discover new relationships by integrating disparate sources; and
- transform these representations to conform to other frameworks of entities and relationships' (Chrisman 1997).

Most definitions in the literature emphasise that GIS is only a tool. One of the common mistakes made in regard to GIS is to expect that GIS will solve spatial problems in isolation. Indeed, this misconception led to a great deal of disillusionment in the 1980s. Stressing that GIS will only *aid* in decision making and problem solving is, therefore, an important concept in defining GIS. Successful application of GIS depends heavily on the user making critical and informed decisions.

Other definitions focus on the purpose for using GIS, and define GIS in terms of its abilities. Additional information regarding the actions supported by GIS further clarifies the usefulness of the tools.

It is helpful to consider each of the acronym letters individually (Figure 1.1):

G: (geographic or geographical) indicates that we are talking about the real, spatial world and considering a quality or quantity that is spatially distributed. The more pedantic would require that this data be registered in recognised coordinate systems, be it a latitude/longitude system, an easting and northing pair, or an *x*- and *y*-coordinate, in order to be 'geographical'.

I: (information) identifies that we have some data (measurements) within the context of a system of meaning. It is from information that knowledge grows. At least one element of GIS information must be linked to the G, i.e. there must be some geographical information (e.g. derived from map coordinates). Other information may be related to attributes (e.g. derived from a name or label), or topology (providing, for example, an understanding of the local environment). Geographical and attribute data types are discussed in Chapter Two.

S: (system) infers that we are relating separate entities using linkages. These entities may be the computer hardware, the software, the data, and the user. When these entities are combined, or linked, they form a system of interactions and interdependencies.

Figure 1.1 Geographical information system

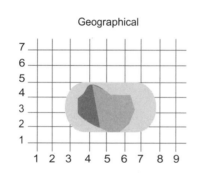

Geographical

Information

Street name = Jumbuck Road

South Australia shares a border with Western Australia

There is a point feature (a well) at coordinate 23, 65

pH = 7

System

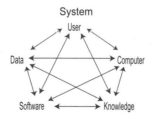

The *Geographical Information Systems: An Introduction* learning experience

This book is structured as a cumulative learning experience. Figure 1.2 shows, diagrammatically, how the reader will progress through this introductory text.

This chapter introduces, describes, and discusses uses for GIS. Chapters Two to Seven are concerned with data: identifying, capturing, editing, outputting and managing it. Chapter Two discusses the data required by the system. Once identified, data must be entered into the GIS. Chapter Three describes this data-entry process. Data entry is rarely error free. Chapter Four discusses the data editing required to remove or minimise these errors. Chapter Five gives a brief introduction to map **projections** as used within GIS. At some stage you, the **GIS user**, will require the system to facilitate **data output**. Data output methods and issues concerning data output are addressed in Chapter Six. As GIS databases become larger, the need to employ good **data management** techniques becomes increasingly obvious. Chapter Seven finalises the discussion of primary concerns with GIS data by introducing data management. With the understanding gained from these chapters you will be prepared to turn your attention to **data manipulation** and analysis.

Chapter Eight introduces the elementary tools available in most GIS software. Beyond the elementary querying and display are the geoprocessing and **overlay** tools, as discussed in Chapters Nine and Ten. These **geoprocessing tools** allow the manipulation of existing data in a manner defined by the user. The concepts and application of overlay, arguably the definitive GIS analysis tool, involving the combination or merging of existing data, are also discussed. Whereas Chapters Eight, Nine, and Ten focus on tools common to most GIS software packages, Chapter Eleven isolates examples of application-specific tools, namely **proximity analysis** and **network** analysis. The knowledge needed to undertake landscape GIS-type studies or analysis, including the construction of three-dimensional **models** and derivatives of elevation data, is supplied in Chapter Twelve. Tools described in Chapters Eight to Twelve can be integrated into GIS-based modelling tools, which offer extremely powerful and flexible modelling environments for spatial phenomena. The types of GIS-based modelling are discussed in Chapter Thirteen.

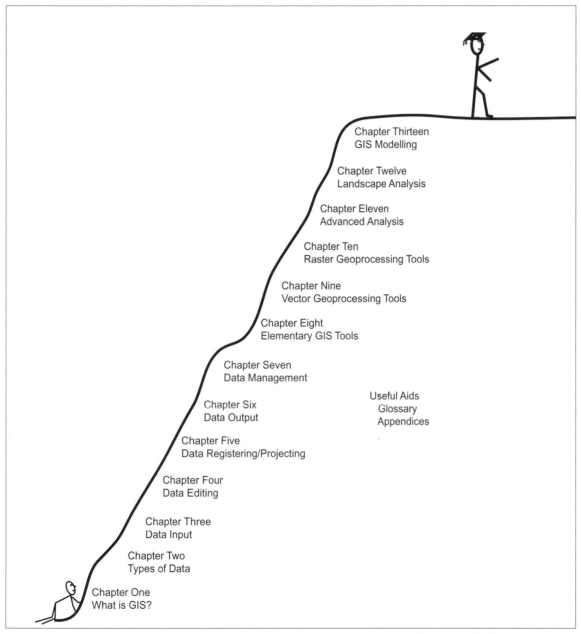

Figure 1.2 Scaling the mountain of GIS understanding

GIS components

The main components of GIS are software, hardware, data, and users (or personnel or **liveware**) (Figure 1.3).

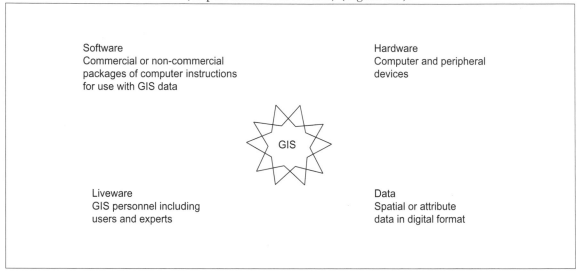

Software
Commercial or non-commercial
packages of computer instructions
for use with GIS data

Hardware
Computer and peripheral
devices

GIS

Liveware
GIS personnel including
users and experts

Data
Spatial or attribute
data in digital format

Figure 1.3 GIS components

GIS software contains instructions to the computer that will be interpreted into actions. One example instruction may be 'Start up the software package', which will result in the GIS software's initial display opening on the computer screen.

There are many GIS software packages, which vary in cost, performance, computing platform, and user-friendliness. Increasingly, these packages are becoming a set of modules that allow the purchaser to select precisely the abilities required for their research or work. For example, it is possible to select a module that produces high-quality graphics, or one that allows the integration of remotely sensed imagery.

The hardware is the actual machinery used by the software. It ranges from personal computers (PCs), to **plotter**s, **digitiser**s and **mainframe**s. A basic system may be expected to have a **visual display unit** (screen or monitor), a **central processing unit** (the actual computer) and a disk drive (for floppy disks or CD-ROMs, as well as an internal hard disk). Other common peripherals include a **mouse**, a **printer** or plotter, and a digitiser (discussed in Chapter Three).

The data fuel the software. These data are usually stored separately from the software. Example data may be census boundaries (spatial data) and census data (attribute or aspatial data).

The user (personnel or liveware) may represent an individual, a department, or an organisation. This is the part of the system that drives all decisions and actions. In order to derive sensible answers from GIS, this user should ideally be knowledgeable in both GIS and the field of interest. For example, a **GIS analyst** with a background in planning and demographics would be an ideal GIS user for an exercise in planning urban expansion zones.

Each of the GIS components relies on and affects the performance of the other components. A slow **machine**, software that requires extensive knowledge of obscure **commands**, old data, or a poor typist are all examples of potential problems in the system.

Functional elements of GIS

The functional elements of GIS relate to its basic duties. These elements provide for **data input**, **data storage**, data management, **data retrieval**, data manipulation, **data analysis** and **data modelling**, **data output** and **data display** (Figure 1.4).

Figure 1.4 GIS functional elements

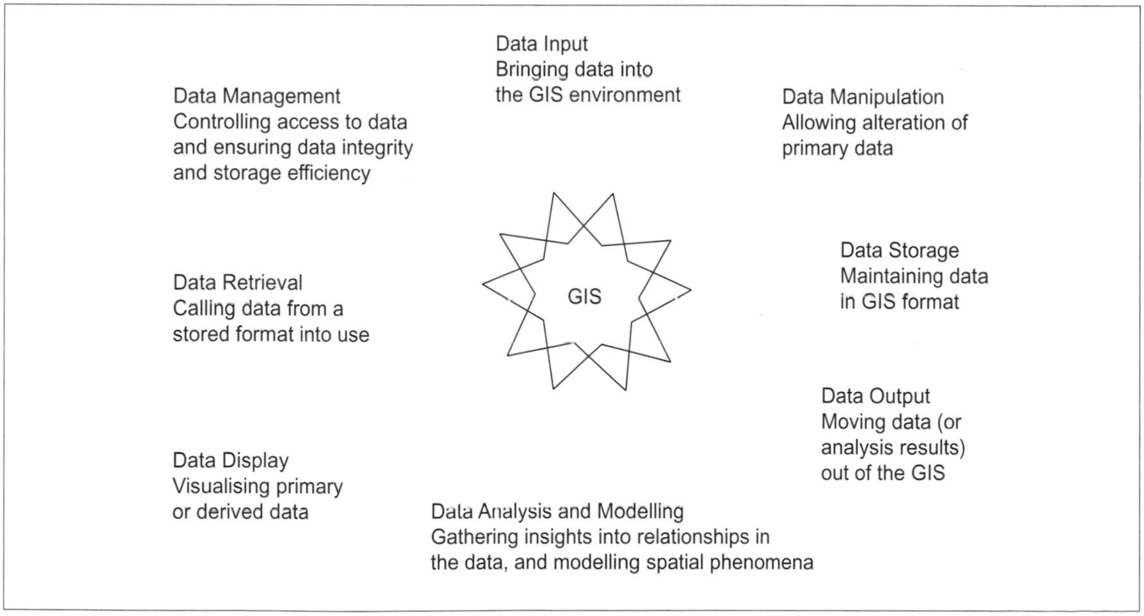

Data Input
Bringing data into
the GIS environment

Data Management
Controlling access to data
and ensuring data integrity
and storage efficiency

Data Manipulation
Allowing alteration of
primary data

Data Storage
Maintaining data
in GIS format

Data Retrieval
Calling data from a
stored format into use

GIS

Data Output
Moving data (or
analysis results)
out of the GIS

Data Display
Visualising primary
or derived data

Data Analysis and Modelling
Gathering insights into relationships in
the data, and modelling spatial phenomena

Data input requires that the GIS user has the tools to gather data from many disparate sources. For example, the GIS database may include data captured from **topographic map**s, satellite imagery, aerial photography, field surveys, other GIS software packages, and spreadsheets. GIS provides a means for collating data. The mechanisms facilitating the capture of data are described in Chapter Three.

GIS store data in a digital format. Storing digital data is remarkably efficient when compared with storing the original products (such as paper maps and survey sheets). The storage method should be reliable and, ideally, take as little space in the computer as possible. GIS therefore offers storage that takes little physical space and is, usually, reliable and secure.

Data management practices are required as GIS databases tend to be large. Managing data may involve tracking data movement through GIS modelling or analysis phases, ensuring data integrity and overseeing security of the data. Proper management of data ensures that storage and access are undertaken in the most efficient manner possible.

To be able to retrieve data is essential for the proper functioning of the system. Users should be able to define which data to retrieve and the manner in which it is to be retrieved. Although this may be a complicated task for the computer, it should always appear simple to the user. This ensures that the user can access data in a timely and defined manner.

Data manipulation and conversion functionalities in GIS allow the user to alter the data. Examples may include changing the map projection, or rotating an image. One of the important roles of data manipulation and conversion is ensuring that data from disparate sources can be compared. This may involve, for example, transforming data referenced with **Map Grid of Australia** (MGA) coordinates to a coordinate scheme of latitudes and longitudes for comparison with other data referenced in this way.

Data analysis is the functionality that clearly distinguishes GIS from **computer cartography**. This form of data manipulation, however, is undertaken specifically to provide new insights into the data and to generate new information and secondary data. Modelling is essentially a combination of a number of different analysis steps, and it is this functionality that has led to GIS being described as intelligent computer cartography or intelligent mapping. It is important to be aware that the primary intelligence lies with the user and not the software or the computer!

Data output in GIS allows digital data to be extracted or exported for use elsewhere. The actual output format may range from a computer file directed to a disk, to an analysis result exported into a spreadsheet. Data output functionality allows GIS personnel to use the data in other software packages or for **hard-copy** (on paper) display.

Data display is really a form of data output. Display may be to a screen (visual display unit), a printer, or a plotter. Display on the screen implies that GIS allows you to see the data, be it in the form of a map, a graph, or a table. Hard-copy output implies that data are printed on paper or film. Hard-copy could also be a map, graph, or table. The effectiveness of the display is an important consideration in GIS. The spatial element often requires that the data be seen in a spatial context, in a manner that eases exploratory data analysis. This may be important in elementary exercises prior to planning data analysis.

How could I apply GIS in my field?

The wide range of functions available within GIS provide useful tools that can be used in many disciplines or fields of study. Although GIS is used most commonly to query databases, arguably its most important or exciting uses are analysis and modelling. The following is a list of traditional questions GIS has been used to solve (modified from Burrough and McDonnell (1998)), and includes some specific examples:

Where is/are _?
    Where is Perth?
    Where are the north facing slopes?
    Show me areas zoned for high-density development.

What is at _?
    What is located at MGA coordinate 345 768, 6 245 785?
    What species of bird would I find at Lake Gwelup?

What is the name of that street?

Where is _ with regards to _?
  Where is the ambulance located in relation to the hospital?
  Where is the quokka living in relation to the fox?
  Are the parks located near the conservation reserves?

How many _ s are there within distance _ of _?
  How many geological faults are within 200 km of the mine?
  How many water troughs are within 1000 m of the flock?
  How many lookouts are within 30 minutes walk of my hotel?

What is the value of function _ at location/s _?
  What is the aspect of the granite outcrop?
  Is the soil erosion risk potential higher in Ararat or Sale?
  What is the cost–benefit ratio for locating a shop here?

How large is _?
  What is the population of Darwin?
  What is the surface area of the water body?
  How large (km^2) is the school student catchment area?

What is the result of intersecting spatial data (maps)?
  Are there any properties for sale in industrial zoned land?
  Do the spatial distributions of frogs and ducks overlap?
  Do all suburbs have similar sized areas of green space?

What is the best path from _ to _ along _?
  What is the quickest way to get from the airport to the conference centre along the road network?
  What is the safest route for the movement of chemical waste from the production site to the waste plant, using non-residential streets?
  What is the cheapest route for a new train track from Port Arthur to Hobart?

How can we reclassify _ and _ into _ and _?
> Can we regroup the data currently showing the spatial distribution of migrants and non-migrants, into people who speak English and those who do not speak English?
> How can we reclassify the spatial distribution of cats and dogs into licensed and unlicensed pets?
> Let us reclassify soils and slopes into ideal planting sites and poor planting sites.

Model process _ over time _ for a given scenario: _.
> Model the effect of logging a forest on soil erosion—from pre-logging to two years post-logging.
> Predict the effect of changing climatic conditions on the distribution of the koala for the year 2050.
> Model the spread of Brisbane's urban sprawl (based on population and housing data from the last ten years) and predict where the edge of the urban region will be in the year 2020.

Many of these questions may be answered using traditional methods, which may take many months, a great deal of patience, and a large amount of money. They are also ideal GIS questions, which can be handled easily and efficiently with GIS technology.

Why use GIS?

There are many reasons for using GIS in tasks ranging from data collection and management to modelling. These include:
- GIS is computer based, and has all the advantages (and disadvantages) of a computer tool;
- GIS is established in many workplaces and has proven not to be a passing fad; and
- GIS can be an accurate, efficient, effective and scientific tool.

As a computer-based tool, GIS is becoming increasingly cost effective, convenient and powerful. GIS can operate on a PC platform, and free GIS software packages exist. It can, therefore, be inexpensive to set up a system, when compared with the costs involved, say, a decade ago. A great deal of data is available in digital format with associated **metadata** so that the user can judge the **accuracy** and reliability of the data.

The computer empowers GIS, and, in turn, the user, with the ability to conduct analysis and modelling exercises. When computer-based GIS is compared with traditional methods—using tracing paper and a number of paper maps—it becomes obvious that GIS has the potential to be more accurate, more precise, more objective and more efficient.

In the 1980s, choosing GIS over traditional methods was fraught with uncertainty. Would GIS perform to the expectations of the user? In some cases expectation exceeded the abilities of GIS and disillusionment followed a large expenditure on data, software, hardware, and personnel. In the majority of cases, GIS became a replacement for computer cartography. Sometimes GIS personnel were unaware of the potential of its analysis and modelling capabilities. Over the past decade there has been growing awareness of the capabilities and potential applications of GIS, and an increase in education and training courses in GIS at all levels from school and college through to tertiary course work and research degrees. There has also been a trend towards affordable, user-friendly software packages. The uncertainty of the past has been replaced with a well-trained or educated work force able to make informed decisions regarding the establishment of a stable and productive GIS environment.

Advances in knowledge and increasing affordability have seen GIS established in larger government departments and in industry. Acceptance at these levels has also led to application of GIS in smaller government departments, local councils and private consultancy firms. The result is that today GIS has a presence in a wide range of fields, and future predictions suggest GIS will become further entrenched in private and public organisations.

GIS has become more popular than the traditional techniques. When the system is operating in the way in which it was intended, it is an ideal tool for use in many tasks, such as decision support, modelling, and even simple mapping. You no longer need to be a draftsperson with a steady hand to make high quality maps. You no longer need a strong mathematical background to undertake spatial statistics.

GIS usage: Comments from the workplace

The functional definition of GIS varies greatly from site to site. Many employers are beginning to realise the potential of GIS; others still define it as a handy cartography tool.

> In most of the positions I have worked in, capture, storage, retrieval, analysis, and manipulation of spatial data formed a major part of my work, but display of this data was the most important from my employer's point of view.
>
> *Narah Stuart, GIS Modeller*

Exercises: Exploring Jarrahlea

Jarrahlea is a small property, approximately 2.4 hectares in area, situated in the picturesque Hills district of Perth, Western Australia. The property experiences Perth's Mediterranean climate of hot, dry summers and warm, wet winters. The main concerns regarding the management of the property are water issues, as the long dry summers almost exhaust the amount of water captured in tanks and dams over winter; and safety issues related to the bush fire season.

The Jarrahlea data set was selected for the development of exercises in this textbook as it is a diverse data set adaptable to a variety of uses. The property has no commercial land use, although comments regarding potential development as a Christmas tree farm, a yabby farm, or a 3-hole golf course have been discussed passionately over cold drinks on a hot afternoon.

The following map (Figure 1.5) will serve as a tool to familiarise you with the property. Only selected data have been displayed in the GIS map hard-copy output. Each different 'layer' of data is represented in the legend. Look at the map and consider the following questions. The answers can be found in Appendix 1.

Q1. How many 'layers' of data are displayed on the map?

Q2. Estimate the area (in square metres) of the property.

Q3. What proportion of all tea-trees are within 10 metres of the creek?

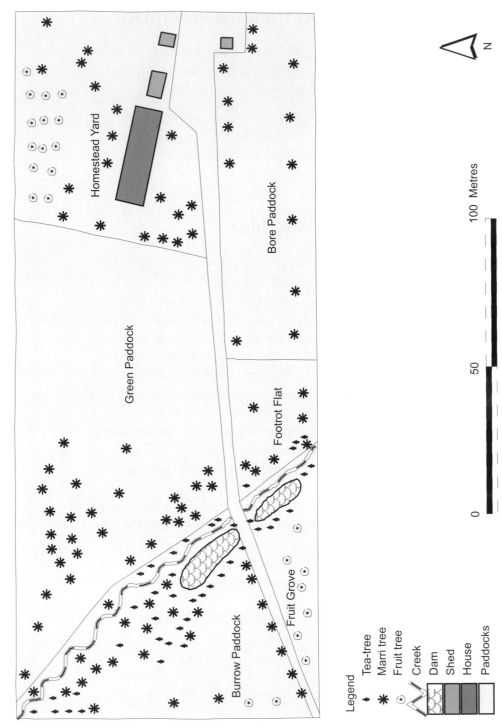

Figure 1.5 Exploring Jarrahlea

Homestead Yard

Green Paddock

Bore Paddock

Footrot Flat

Burrow Paddock

Fruit Grove

Legend

- ➤ Tea-tree
- ✳ Marri tree
- (•) Fruit tree
- Creek
- Dam
- Shed
- House
- Paddocks

N

0   50   100 Metres

CHAPTER TWO:
Data

This chapter presents an introduction to the array of data that constitutes the digital data in GIS. After completing this chapter, the reader should be able to identify the two data types—spatial and attribute—and should understand the concept of topology. GIS integrates data types (vector and raster, spatial and attribute data) and data connectivity (topology) to create an ideal analysis and modelling environment for geographical data.

The types of data

Data is usually the driving force of a GIS project. Most GIS are designed to deal with two types of data (Figure 2.1): spatial data (Where is this feature?) and attribute data (What is this feature?). A third element, **topological** understanding (What surrounds this feature?), is constructed in order to describe connectivity between spatial features. In GIS these two data types and the topological element are intrinsically interrelated. They may be entered and stored separately, however, the ability to have the three simultaneously in the same computing environment aids the high-level analytical abilities of the GIS and allows the GIS user to manipulate data to produce new information.

Spatial data: Where is it?

Spatial data locate features in space. Spatial data are normally expressed as real numbers (numbers with digits after the decimal place, such as 23.543) or as integers (whole numbers, such as 24). Examples of spatial data include:

- a latitude–longitude coordinate: 37 degrees, 0 minutes South and 149 degrees, 43 minutes East (37.0S 149.72E)
- a **grid cell** location: row 3 and column 5; and
- 10.564 kilometres NW of the Black Stump.

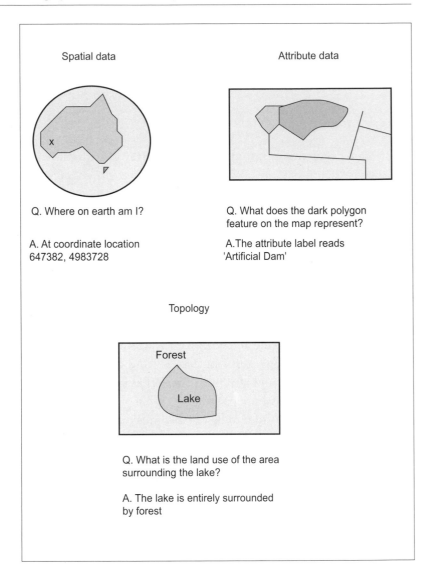

Spatial data

Q. Where on earth am I?

A. At coordinate location
647382, 4983728

Attribute data

Q. What does the dark polygon
feature on the map represent?

A.The attribute label reads
'Artificial Dam'

Topology

Forest

Lake

Q. What is the land use of the area
surrounding the lake?

A. The lake is entirely surrounded
by forest

Figure 2.1 Types of data

Attribute data: What is it?

Attribute data describe a feature at a spatial location. Attribute data
may be text strings (words) or numbers. Text string, or word, attributes are commonly nominal data and are usually represented by
names, such as 'owner: Jones, Smith, or Jenkins'. Nominal data simply indicates what to call the object. Numerical attributes may be
real or integer numbers, Boolean (0 or 1), or ordinal, ranked data,
such as low, moderate, and high slope. The choice of how to repre-

sent attribute data will affect the ability to use the data. For example, nominal data are not used with numerical operators.

Example attribute data may be:

- the population is 2513 people
- the species is *Eucalyptus sieberi*
- asthma likelihood is 0.92.

Topology: What is its environment?

Topology embodies spatial data relationships. This element can empower the GIS in many ways. It ensures that data 'knows' what is where. It links the attribute and spatial data and understands relationships with surrounding data, i.e. is aware of the neighbouring spatial and, therefore, attribute data. This knowledge is important in modelling and analysis. Example GIS queries that require topological information are:

- Can I access Karrinyup Place from Gwelup Road?
- Do young families tend to live next door to other young families?
- Does the eucalypt forest border on the pine forest?

Some GIS software packages require topology to be 'built' by the user issuing a command. In other packages, the topology is created automatically as spatial and attribute data are added and linked.

Data structures

Data is usually stored in one of two **data structure**s—vector or raster (Figure 2.2). The form used will affect the storage and display of the data as well as the type of analysis and manipulation the user can perform.

Vector data based GIS is perhaps the easier structure to understand for the non-computer scientist. The vector world appears as we are used to seeing it on a map. Entities can be **polygon**s, or areas (e.g. a Local Government Area), **line**s (e.g. a road network), or **point**s (such as a well location). Lines are composed of end points (nodes) and inflection points between **node**s (vertices). Polygons, lines and points are sometimes called geographical primitives as they represent the smallest units of spatial information in a GIS layer of data. The user's choice of the correct geographical primitive to represent data

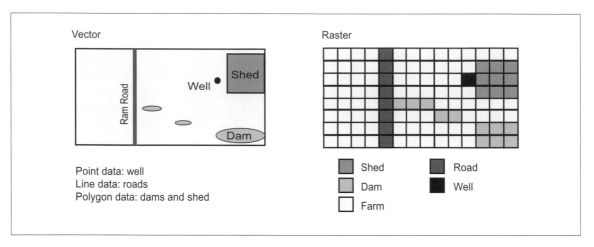

Figure 2.2 Data structures

will depend on **scale**, or the desired level of generalisation in the data. A city may be best represented by a dot on a small scale map (say 1:1 000 000), whereas it may be best represented by a polygon on a large scale map (say 1:1000). Vector GIS is capable of being more accurate than raster GIS in defining spatial location, although it is generally more expensive in terms of data storage and computing power. Vector GIS is able to support topology explicitly.

Raster data refers to a world that is divided into cells. The geographical primitive of raster data is the cell. Cells are usually square and contain one number or key code, which links to an explanation (for example, 1 is a lake, 2 is a forest) and a shading scheme. Raster is a relatively easy way of storing and using data, and complements computer-based technology. Computer screens, for example, use raster-type displays. Another advantage of raster structures is that they are compatible with remotely sensed imagery, such as imagery captured by detectors on satellites, and any data source that shows data varying continuously over a surface (i.e. **continuous data** rather than **discrete data**). Raster structures are lower in spatial accuracy than vector structures, as the smallest entity is always a cell, which represents an area and not a coordinate. Raster data may be lower in attribute accuracy also, as cells usually represent the feature that fills most of the cell. Determining the cell size in a raster display will be a crucial step in assembling a useful data set. Finally, raster does not support topology explicitly, rather it allows topology to be recognised implicitly.

Software packages may support one or both of the data structures. Conversion between the two data structures may be undertaken with varying levels of success (Figure 2.3). Many software packages will allow the user to 'smooth' vectorised raster data and a small cell size can give a visually pleasing raster line.

Figure 2.3 Conversion between data structures

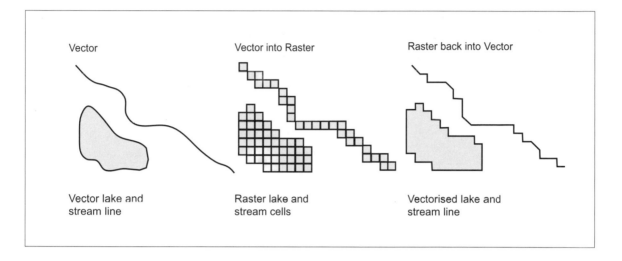

Data: Comments from the workplace

Many divisions or departments in Australia using GIS initially drew on data from Computer Aided Design products. This meant that the spatial element and the attribute element of the data were often divorced, or stored separately, and topology was non-existent. This problem was mentioned by a number of survey respondents. Most data providers have made, or are planning to make, the transition to providing data in a GIS format in order to meet the demands of clients.

Another data-based problem identified was that of attempting to communicate data requirements to data providers not conversant with GIS terminology or requirements.

> Data is everyone's biggest headache. The problem is often that the data
> providers have little idea of the requirements of a useful GIS dataset.
> *Matthew Aylward, Part-time GIS Operator*

Exercises: Jarrahlea data recognition

Examine Figures 2.4 and 2.5, and answer the questions provided below.

Q1. Figure 2.4 displays vector data. Can you identify and name the point layers, the line layers, and the polygon layers?

Q2. How does scale (the level of generalisation) affect the manner in which you represent vector data?

Q3. Figure 2.5 displays the same data as Figure 2.4. Two of the layers have been altered. One layer has been converted to a raster **grid cell layer**. Another layer has been converted to a raster grid cell layer and then re-vectorised and 'smoothed'. Can you identify these altered layers?

Q4. Identify the spatial, attribute, and topology elements in the following statement: The paddock called Footrot Flat (centred approximately 95 metres south-west of the homestead) is adjacent to the western boundary of Bore Paddock.

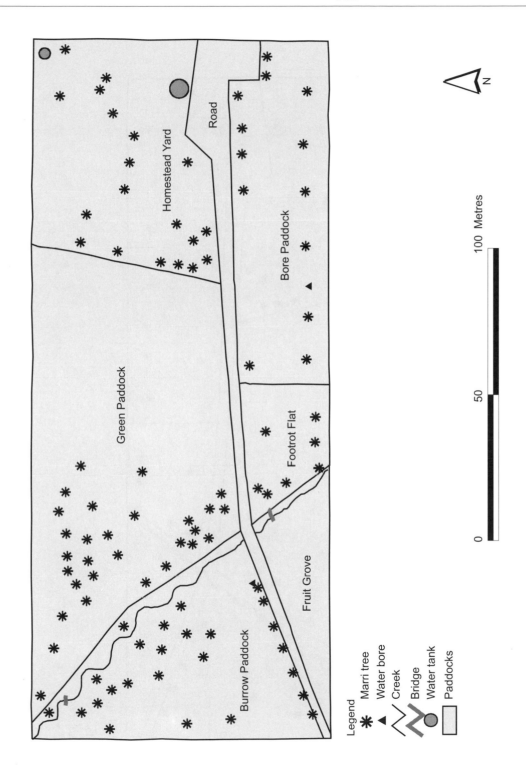

Figure 2.4 Jarrahlea data recognition

Legend

✳ Marri tree
▲ Water bore
〰 Creek
〰 Bridge
● Water tank
▢ Paddocks

0   50   100 Metres

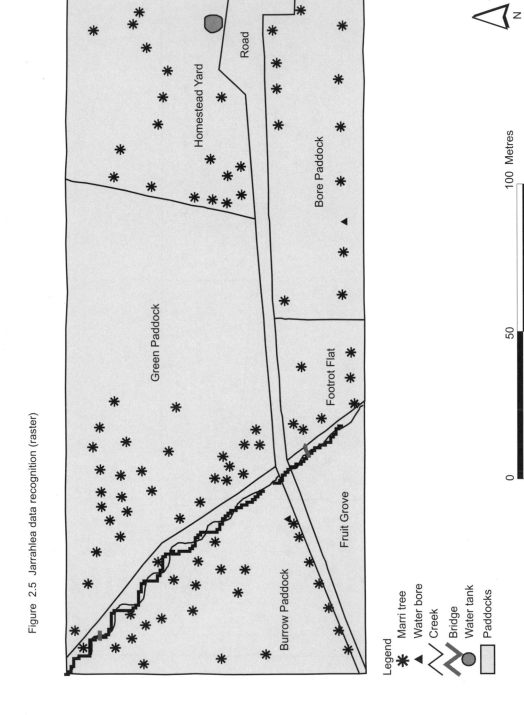

Figure 2.5 Jarrahlea data recognition (raster)

Legend

Marri tree

Water bore

Creek

Bridge

Water tank

Paddocks

100 Metres

50

0

N

Homestead Yard

Road

Green Paddock

Bore Paddock

Footrot Flat

Burrow Paddock

Fruit Grove

CHAPTER THREE:
Data Input

Data input, the initial stage in any GIS project, is a complex and expensive process that requires much consideration. The GIS user needs to know which software and hardware they are going to be able to access throughout the life of the project, the most appropriate format and structure of the data, and the requirements of the project.

Realising that data input can be the most expensive and time-consuming aspect in GIS should stimulate sufficient interest in trying to streamline the process.

Data input

Gathering and converting data into digital formats (the process of data input) is notoriously expensive and often tedious and time consuming. A complicating factor is that GIS has two types of data and these forms of data may need to be entered separately.

Data input methods

There are various data input methods (Figure 3.1).

The most common are:

- *digitising*: This is a process whereby the user traces line work on a map using a puck, which looks very much like a computer mouse. The tracing takes place on top of a digitising board that has a mesh of sensory wires just beneath the surface. When the digitising puck button is depressed the wire mesh registers where the puck is located and stores this location in digital format. Thus paper maps are digitally 'traced'. This input method is used primarily for spatial data. It is, however, possible to create an attribute key, or legend, on paper and place this on the digitising board to facilitate digitiser entry of attribute data. The user may trace a feature (say, a road) using the puck and then click on a box in the key that has been identified as a certain attribute number or class (say, highway). In this manner spatial and attribute data are input with the one process.

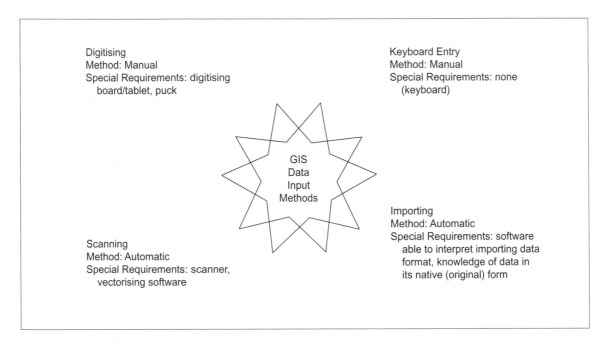

Figure 3.1 The four main data input methods

- *keyboard entry*: This involves typing data using the keyboard. It is common to type in attribute data and even spatial data for a data-sparse map. Most users, however, will avoid keyboard data entry if possible. Not only is it time consuming and tedious, it is also error prone for non-typists.

- *scanning*: This process is also termed **automated digitising**. This method uses a device (scanner) that passes a photoelectric cell or 'eye' over a paper document and records lines, points, text, and any mark on the data source. It functions much like a paper-to-digital photocopier. In many cases, scanning technology has not been readily accepted due to the high proportion of time required to edit and 'clean' the scanned image. The scanner is a 'dumb' data gatherer. Text, symbols, annotation, and smudges or coffee cup rings are treated as equally important pieces of line work. An ideal scanning document would be a white piece of paper with distinct black line work of one feature only (such as a road line). Success using this technique depends on the scanner **resolution** and the abilities of the **vectorising** software (when appropriate).

- *importing*: Importing the data implies that the data is already in digital format. It may or may not be in the correct format for your GIS software and hardware. Importing data can be either the easiest or the most frustrating technique for data entry, depending on the data formats and software involved.
- ***methods using other technological tools***: A prime example would be the use of a Global Positioning System (GPS).

Issues of data input

The main issues that should be considered prior to determining a method for data input include (Figure 3.2):

- the balance required between time and effort spent on data capture and editing errors resulting from the technique used;
- the cost of the capture process;

Figure 3.2 Issues and considerations concerning data input

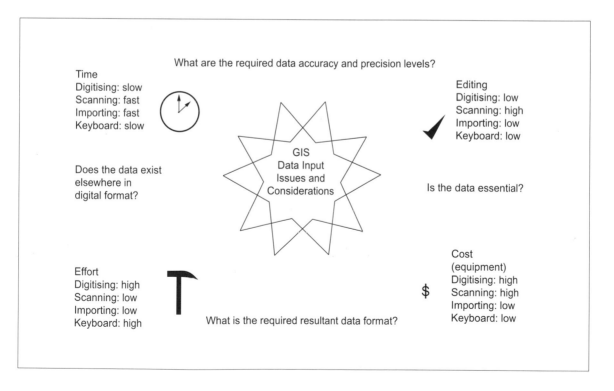

- the necessary format of the resulting data;
- the required accuracy and **precision** of the data; and
- whether a similar data input process has already been undertaken elsewhere.

Generally, the quicker the data input (capture) phase, the longer the editing phase. Scanning make take seconds; however, the editing of this scanned data make take weeks. Accurately digitising may be slow and arduous, but offers the reward of requiring minimal editing in comparison with scanning. Keyboard entry is slow and may be error prone; however, typing in the spatial coordinates may allow a level of precision which cannot be obtained using a mouse, puck, or scanner.

The cost of data capture is a major concern, as data is typically expensive. In many cases the actual cost of the software and hardware pales into insignificance when considering the costs of gathering data and the hours spent putting the data into the system, or the purchase of pre-existing digital data. Quality data entry equipment, such as digitisers and scanners, are expensive. As a result, outsourcing data input is quite common. Many data sets, however, must be dynamic in order to be relevant. These are data sets that require continual and regular updating, such as **cadastral** boundaries. In this situation, investing in high quality equipment that offers efficient means of data capture is money well invested.

The format of the data may determine which input method is appropriate. Data that are inherently raster, are easily scanned. Data supplied as paper lists of coordinate pairs with attributes recommend keyboard entry. Complex paper maps, such as topographic maps, are prime candidates for digitising.

Accuracy and precision are two concepts that can be crucial to any GIS data input project (Figure 3.3). Accuracy refers to the degree of error between an estimated or modelled location and the actual location. Levels of precision are based on the level of detail in the data definition, e.g. the number of digits after the decimal point. Therefore it is possible to be very accurate and imprecise or very precise and highly inaccurate. Digitisers and scanners can be very precise but inaccurate if not calibrated correctly, or if the equipment is outmoded. Keyboard entry allows the user to define the precision and maintain a set level of accuracy.

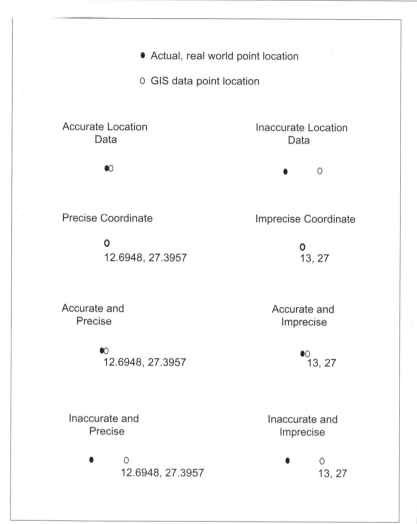

Figure 3.3 Accuracy and precision

The final factor to consider before investing in a lengthy and expensive data capture phase is the availability elsewhere of similar data in a useable format. Redigitising a contour map, which can be bought as a digital data product from the data custodian, would not be cost effective or time efficient. When identifying others who have digital versions of data that are of interest to a project, it is crucial to determine that the pre-existing data is accurate enough and timely enough for the purpose at hand.

As an aside, it is interesting to consider the idea that data, once captured, may be sold to offset the capture costs. This is practised by a number of users of spatial data. Issues such as freedom of access to information and copyright laws regarding data should also be considered.

Data input: Comments from the workplace

Australia has a data-hungry GIS community. In many situations, large data input exercises have been well planned and well funded. Many base data sets are now established and regularly updated. The recognition of alternative uses for GIS, however, will always identify new data to be captured, and may lead to existing data being captured at increasingly higher levels of resolution (larger scales) or with higher levels of accuracy and precision. Although the initial flurry of labour-intensive data entry has subsided, undergraduate GIS students still seem able to find a few hours of digitising in a contractual role to help pay the bills.

A commentary on data input, particularly importing data:

> I try to avoid data input as much as possible! Within Australia it is possible to do this. Within the last year I have discovered the use of spreadsheets for data input, especially when dealing with point data or adding data to existing data sets. Most GIS software will read [spreadsheet] files, which are easily generated from spreadsheets. If you can obtain spreadsheets or (spreadsheet files) from the original data custodian, this could reduce data entry errors. I find attribute data entry using the available methods in most GIS rather tedious.
>
> *Narah Stuart, GIS Modeller*

This is typical of responses in the survey. Most GIS users will pursue all possible avenues to gather data already in a digital format before subjecting themselves, or others, to lengthy digitising or keyboard entry sessions.

Exercises: Capturing the Jarrahlea digital data

Figure 3.4 displays example layers of data, a summary of sources of data and information, and a list of data input methods. Develop a

Figure 3.4 Capturing the Jarrahlea digital data

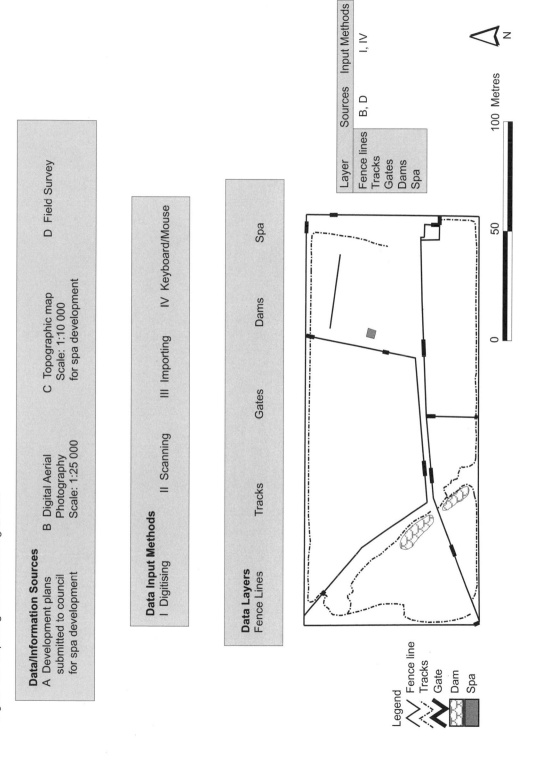

Data/Information Sources

A Development plans submitted to council for spa development

B Digital Aerial Photography Scale: 1:25 000

C Topographic map Scale: 1:10 000 for spa development

D Field Survey

Data Input Methods

I Digitising   II Scanning   III Importing   IV Keyboard/Mouse

Data Layers

Fence Lines   Tracks   Gates   Dams   Spa

Layer	Sources	Input Methods
Fence lines	B, D	I, IV
Tracks		
Gates		
Dams		
Spa		

N

0   50   100 Metres

Legend

Fence line
Tracks
Gate
Dam
Spa

data input scenario for the creation of the Jarrahlea digital data. Be aware that any source may lead to more than one layer and similar methods may be used for many layers. The first data input path has been established for you.

Q1. Describe a possible data input path used to create the fence line layer.
Layer = fence line
Source = aerial photography, field survey
Data input technique = digitising, keyboard/mouse

Q2. Describe a possible data input path used to create the track layer.
Layer = track
Source =
Data input technique =

Q3. Describe a possible data input path used to create the gate layer.
Layer = gate
Source =
Data input technique =

Q4. Describe a possible data input path used to create the dam layer.
Layer = dam
Source =
Data input technique =

Q5. Describe a possible data input path used to create the spa layer.
Layer = spa
Source =
Data input technique =

CHAPTER FOUR:
Data Editing

This chapter discusses errors resulting from the data input stages. Errors should be corrected prior to any further processing or analysis of data. Finding the errors may involve simply looking at the data on the screen in map format, listing attributes, or it may require a more intensive search using other GIS tools.

Data editing

No matter how careful the user and how sensitive the equipment, errors in data entry are almost unavoidable. As data may be vector or raster, and spatial or attribute in nature, there are a variety of data entry methods and a variety of possible associated errors that require editing. The ease of the editing process will depend on the software editing tools.

Data entry errors

Errors can relate to the entry of the spatial data or the attribute data, and can become obvious during or after the creation of topology.

Errors in the spatial data can be divided into errors related to either the vector or the raster data structures (Figure 4.1). In vector GIS, points may have incorrect coordinates due to typing mistakes or incorrect puck or mouse positioning. Lines, being complexes of points, may have node (line end points) or **vertex** (mid-line inflection points) errors for similar reasons. Polygon errors include dangles (overshoots and undershoots), weird polygons (enclosed regions which are not evident in the original data), and slivers (due to digitising a polygon twice, either entirely or partially). As spatial location in raster GIS is referenced to the centre of a cell, or a cell corner, spatial errors will be incorrectly geo-referenced cells.

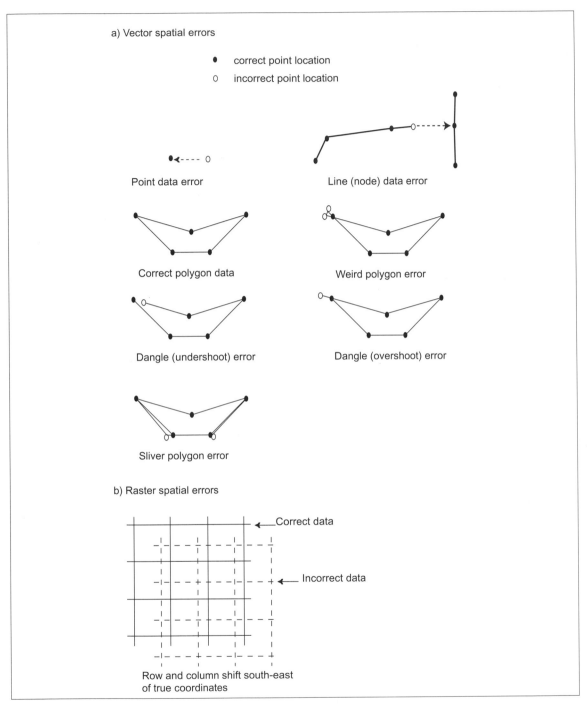

a) Vector spatial errors

● correct point location

0 incorrect point location

Point data error

Line (node) data error

Correct polygon data

Weird polygon error

Dangle (undershoot) error

Dangle (overshoot) error

Sliver polygon error

b) Raster spatial errors

Correct data

Incorrect data

Row and column shift south-east
of true coordinates

Figure 4.1 Data entry spatial errors

Errors such as these originate when the grid location is defined either prior to creation or in registration (Chapter Five).

Attribute errors are incorrect labels attached to spatial data (Figure 4.2). These may occur because the user has been inconsistent in the data entry or has not complied with standard rules for data definitions. For example, 'Low Erosion Risk' is not the same as a category 'low erosion risk' due to the change in capitalisation. Other possible error sources include typing errors. Raster data attribute errors, or misrepresentations, result from the loss of attribute data due to generalisation or an inappropriate raster cell size.

Topology is usually built during or on completion of data entry (either for an entire layer or for one spatial element). If a layer has spatial errors these can lead to attribute or topology errors (Figure 4.3). An example would be two neighbouring polygons, one attribute 'A', the other 'B', with an incomplete dividing line. When topology is built, the software will only assign one label per polygon. The software searches for an enclosed area and in this example 'A' and 'B' form only one completely enclosed area. This is assigned one of the attribute labels ('A' in this case).

It is important to realise that there are many other potential sources of error in GIS. Each phase has a collection of possible pitfalls and the user must be constantly aware of the hazards. Even more alarming is the fact that a small error in an initial stage, such as in data entry, may escalate through the process of a model and become a significant, glaring error in the output. This reinforces the fact that GIS is only a tool and must be managed and used by an intelligent and informed user in order to produce reliable results.

Tools for correcting errors

Correcting data input errors may be a manual or automatic process.

Manually correcting mistakes usually implies moving points or line work using a mouse, or retyping attribute data. Depending on your data input method, data volume, and data type, this may seem a daunting prospect.

Automatically correcting errors is highly attractive in terms of the time required for the editing process; however, it is in itself error prone. Automatic correction is often applied to polygon dangles. An

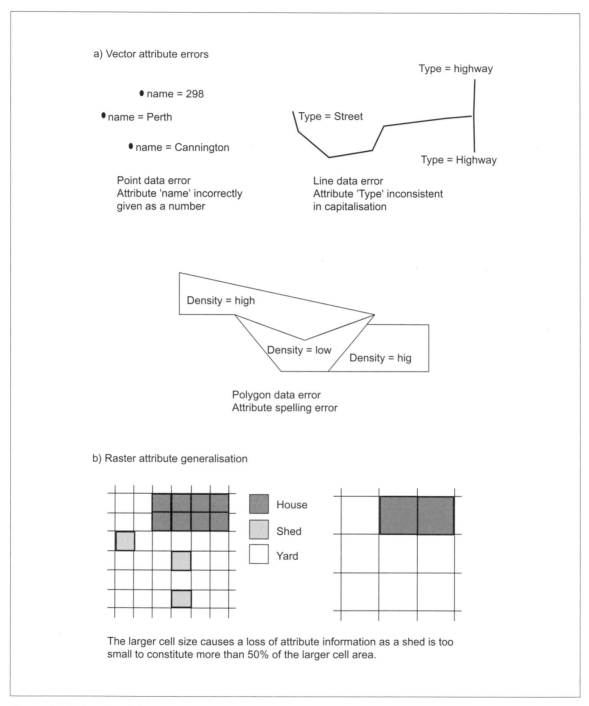

a) Vector attribute errors

● name = 298

● name = Perth

● name = Cannington

Point data error
Attribute 'name' incorrectly
given as a number

Type = highway

Type = Street

Type = Highway

Line data error
Attribute 'Type' inconsistent
in capitalisation

Density = high

Density = low

Density = hig

Polygon data error
Attribute spelling error

b) Raster attribute generalisation

House

Shed

Yard

The larger cell size causes a loss of attribute information as a shed is too
small to constitute more than 50% of the larger cell area.

Figure 4.2 Data entry attribute errors

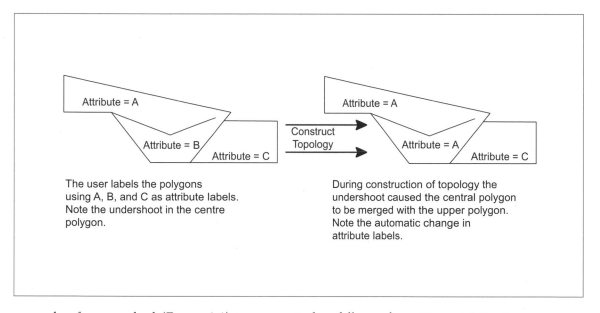

The user labels the polygons using A, B, and C as attribute labels. Note the undershoot in the centre polygon.

During construction of topology the undershoot caused the central polygon to be merged with the upper polygon. Note the automatic change in attribute labels.

example of one method (Figure 4.4) is summarised as follows: the user gives the software a threshold distance, say 10 metres. The software locates the first dangle and then searches for another dangle within a circular search field with a 10-metre radius, with the original dangle as the circle centre. If one or more dangles or nodes (ends of lines) are found, the closest one is joined to the circle centre dangle, based on a user-specified method (move only first dangle, move only second dangle, or move both dangles an equal distance). Other automatic editing examples include conducting searches for small polygons, or polygons with a small area-to-perimeter ratio, identifying them as slivers, and deleting the possibly inappropriate line work.

Figure 4.3 Topology errors

Why is this seemingly attractive data editing tool to be used with caution? An inappropriate threshold distance can cause the data to be oversimplified or rendered useless. Act in haste and be prepared to spend many hours recapturing the data or correcting newly generated errors. The golden rules are:

- get to know the data as well as possible. View the data, and measure appropriate search distances (using the GIS). It is also necessary to read any available metadata, which is data about the data, paying particular attention to items such as scale and accuracy; and
- keep a backup version (copy) of the original, unedited data.

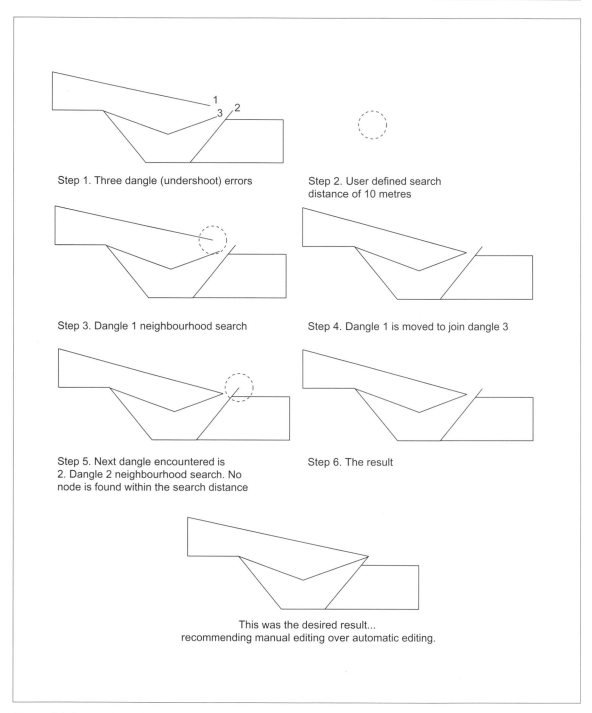

Step 1. Three dangle (undershoot) errors

Step 2. User defined search distance of 10 metres

Step 3. Dangle 1 neighbourhood search

Step 4. Dangle 1 is moved to join dangle 3

Step 5. Next dangle encountered is 2. Dangle 2 neighbourhood search. No node is found within the search distance

Step 6. The result

This was the desired result... recommending manual editing over automatic editing.

Figure 4.4 Automatic spatial error correction

Many software packages have topology building tools which incorporate editing tools. For example, while building topology, the user may define a search radius for joining dangles.

Tools used to correct raster errors require a degree of user input. The most common complaint is loss of information due to over-generalisation (Figure 4.2). The best way to overcome this may be to recreate the grid from the original data source using a smaller cell size. This may not always be appropriate or possible.

Data editing: Comments from the workplace

Many GIS users do not have a background in information technology and may not be equipped with good database design skills. This has led to databases being established without consideration of strict attribute definitions, thus encouraging attribute errors. This may well be a problem of the past as many GIS application courses do now include topics such as database design.

> After a while you learn not to assume anything about the data you have except that it's going to have problems. Sometimes you're nicely surprised. I've stumbled on datasets that have digits dropped off coordinates, that have had attributes stripped to save space and that are in incorrect projections.
>
> *Piers Higgs, GIS Consultant*

Survey comments regarding successful error detection and removal suggest that identifying errors takes time and careful investigation. These are skills that are acquired with practice. The GIS user should rapidly develop a technique for efficient error checking and editing.

> There is a standard set of data errors that cause 95% of all problems. After a while they become relatively easy to spot. The remaining 5%, of course, are the hard ones that keep the GIS operator on his [or her] toes.
>
> *Werner Runge, GIS Project Officer and Consultant*

Data editing involving imported data has plagued many a new GIS user. Much of the older data used in Australian GIS were data originally digitised within a Computer Aided Design (CAD) package rather than a GIS. CAD packages are not fussy about dangles, as they do not build topology. As a result, lines will not meet where the user would expect them to meet and polygons will not exist. In this situation the editing phases may be long and complicated.

> Editing contour data that has been converted [from a CAD package] is nightmarish.
>
> *Narah Stuart, GIS Modeller*

Exercises: When Jarrahlea is not as it should be

Detecting errors in GIS data is a skill that requires practice. The following exercise will test your skill in determining possible errors in a data set. Some of the Jarrahlea layers are shown in Figure 4.5. A table of fruit tree attribute data is also given. The record number, shown in the table, is a unique record identifier assigned by the GIS.

Q1. Study the figure and table to see how many errors you can identify. The errors could be grouped into spatial, attribute, and topology errors. Consider how these errors may cause problems in any GIS analysis undertaken with this data.

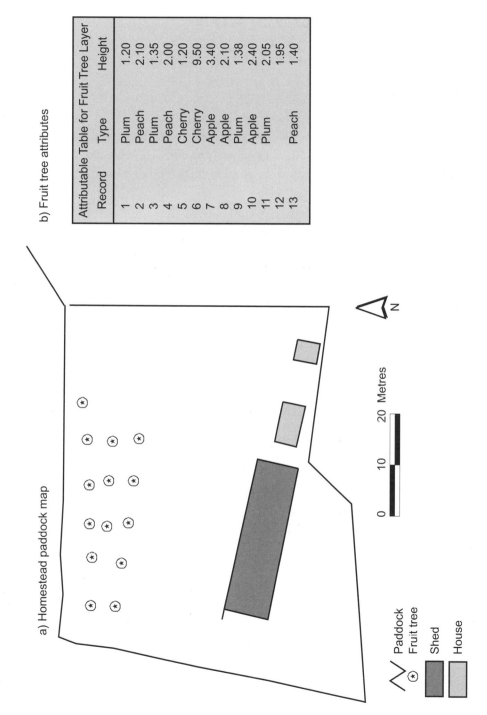

Figure 4.5 When Jarrahlea is not as it should be

a) Homestead paddock map

b) Fruit tree attributes

Attributable Table for Fruit Tree Layer

Record	Type	Height
1	Plum	1.20
2	Peach	2.10
3	Plum	1.35
4	Peach	2.00
5	Cherry	1.20
6	Cherry	9.50
7	Apple	3.40
8	Apple	2.10
9	Plum	1.38
10	Apple	2.40
11	Plum	2.05
12		1.95
13	Peach	1.40

N

0   10   20 Metres

Paddock
Fruit tree
Shed
House

CHAPTER FIVE:
The G in GIS

Following data input and editing, data should be registered into a coordinate system and projected. These processes allow the data to be aligned with data sets of the same or neighbouring areas, and enable the data to be located in the geographical space of the real world.

Putting the G in GIS

Data input is only stage one of the GIS database building process. The next stage involves linking the data in digital space to real world space. The way in which this process is undertaken may depend upon the data-entry method and the data themselves. Generally, the stage following input is a two-phase process that involves registering and projecting digital data.

Artificial coordinate systems

The process of capturing data in digital format often happens with little regard for the actual geographical referencing of the data. The extent to which this statement is true will vary between different GIS packages. When using a digitiser board and the digitising module of a GIS software package, an artificial grid system measures the distance a point is from an origin. The measure is given as two distances, along the **x-axis** and up the **y-axis** (Figure 5.1). The divisions on the axes may be millimetres, or similar units, and the origin is generally the lower left-hand corner of the digitising space. These units are called **digitiser unit**s.

An alternative artificial system uses **screen unit**s (Figure 5.1). These units are important in **heads-up digitising**, the name given to digitising using a mouse on a mouse pad and the screen display. Distance is measured relative to the position of the data on the screen. The origin is usually the lower left-hand corner of the display **window**. These coordinates are only useful at the time they are read. A change in window size or **map extent** will alter the coordinate values.

Figure 5.1 Artificial coordinate systems

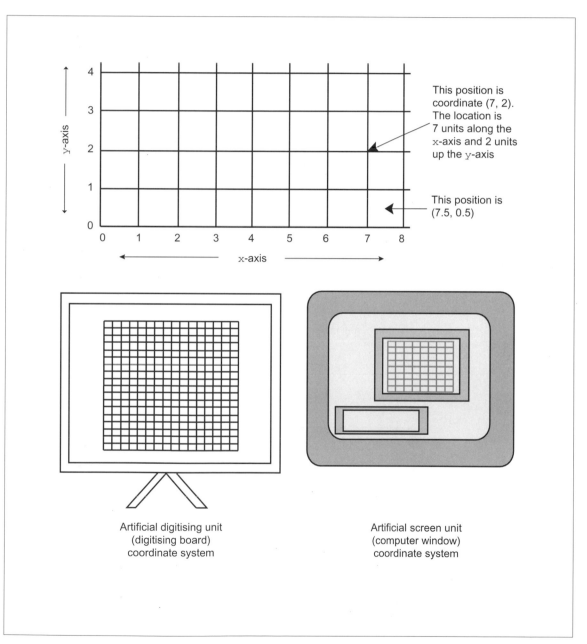

This position is coordinate (7, 2). The location is 7 units along the x-axis and 2 units up the y-axis

This position is (7.5, 0.5)

Artificial digitising unit (digitising board) coordinate system

Artificial screen unit (computer window) coordinate system

Real coordinate systems

To register data to a real rather than an artificial coordinate grid, the user must convert the data to, or align the data with, a standard coordinate system. This process is called registering or geo-referencing (Figure 5.2).

Figure 5.2 Registering data

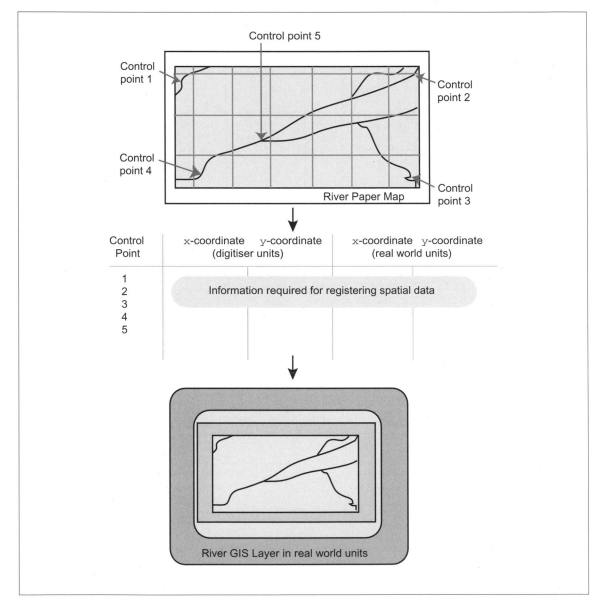

To undertake this process requires the establishing of links between the real world and the digital data. These links are termed **control point**s. A control point is a location for which the real world coordinates are known and the digital coordinates can be read easily. Most software packages require the user to indicate control points (by clicking on the points with the puck) before inputting data using a digitiser. This click on the digitiser indicates where the control points are spatially located relative to one another and the artificial grid. The actual (real world) coordinates may or may not be entered with the control point spatial coordinate. If the real world map coordinate is typed in, the user avoids having to deal with digitiser units and the software will do all transformations without instruction. Alternatively, actual real world coordinates can be entered when the data input stage has been completed. Actual real world control point coordinates—Map Grid of Australia coordinates, Latitude–Longitude, or any other system—are usually entered from the keyboard and the spatial data entered is transformed (scaled and shifted) into this system. Most software packages will do this operation in a manner that appears to be very simple, rather than burdening the user with the complexity of the mathematical algorithms.

How are control points selected? They should be features that are obvious on the map and in the digital data. Road intersections, trigonometric stations, and landmarks are popular choices.

How many control points are necessary? Most software packages will require at least four control points. Generally, the more control points the better the transformation (i.e. the lower the distortion). Four control points are satisfactory, if they are widely spaced. It is common to locate one control point towards each of the corners of a rectangular study area and a fifth, if possible, in a central location.

Why register data? It is the registration process that allows the data to be related to the real world. By putting the data into a recognised coordinate system, the user enables the basic GIS analysis of overlay. This also allows for the correction of any errors, such as distortion due to photocopying, that may have been inherent in the original map.

Projecting data

All geographical data viewed as a two dimensional (2D) array are distorted. The distortion results from that fact that the earth is curved around a sphere in three dimensions (3D). Map projections allow the three dimensional sphere to be viewed, or projected, onto a two dimensional surface, such as a screen or a piece of paper.

There are a vast array of projections, such as Mercator, Robinson and Sinusoidal. Each projection allows the viewing of 3D data as 2D data by a unique approximation. As a result, they all distort a certain feature of the data; for example, some projections distort area and others distort shape.

The initial projection for GIS data should be the projection of the original data source. This is usually given in the map annotation on a paper map. Once a projection has been defined, many GIS software packages will allow transformation between different projections. This allows the user to conduct analyses using the most appropriate projection, i.e. the one that has the least possible distortion for the factor of interest.

Registration and projection: Comments from the workplace

In Australia most GIS users employ, or will shortly make the transition to, the datum called the Geocentric Datum of Australia, GDA, to compute the geocentric coordinate, GDA94. The '94' indicates that the datum used as the basis for the coordinate system was correct on the 1 January, 1994. Datum change location slowly over time due to natural phenomena such as plate tectonics. The actual coordinates are called Map Grid of Australia 1994, MGA94. MGA94 is a localised, Australian version of the Universal Transverse Mercator.

There are uses of alternative systems for projects that are defined by a greater expanse than the Australian continent. However, MGA94 (and its predecessor, Australian Map Grid) is the most widely accepted scheme in Australia. Wide acceptance of MGA94 facilitates easy comparison of data gathered from a number of sources.

We are lucky with [Map Grid of Australia] projections. In south-east Asia, where I worked briefly, I saw more projections used than I had in a year at home.

Piers Higgs, GIS Consultant

Registering or (georeferencing) and projecting data are difficult concepts to understand for those without a strong cartography, surveying, or mathematical background. GIS software has been written to make these complicated transformations easy to use. This is both potentially advantageous, to those who know enough to make sensible decisions, and distinctly dangerous to those who guess. It is good advice to seek help from someone who is an expert in these fields, as suggested by the comment below.

I find projections a bit confusing, and often seek the assistance of a colleague when having to deal with them.

Narah Stuart, GIS Modeller

Exercises: Coordinating Jarrahlea data

Figure 5.3 shows three example maps of the Jarrahlea property. The first (Figure 5.3a) is given in an artificial coordinate system, the second (Figure 5.3b) is given in real world measurements which are not based on a coordinate system, while the third (Figure 5.3c) shows the property with MGA94 unit control points.

Q1. Do you think that the artificial scheme in (a) represents digitiser units or screen units?

Q2. Why would a GIS user employ the units shown in (b)?

Q3. What does 1 unit in MGA94 coordinates represent in real world coordinates?

3D Australia in 2D

Figure 5.4a–c shows maps of Australia in differing projections. Australia is used as an example, as the distortion may not readily be obvious on a property the size of Jarrahlea. Distortions are, however,

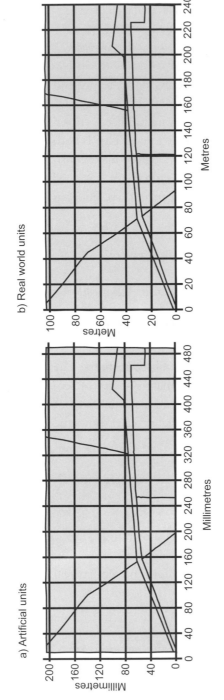

Figure 5.3 Coordinating Jarrahlea data

a) Artificial units

b) Real world units

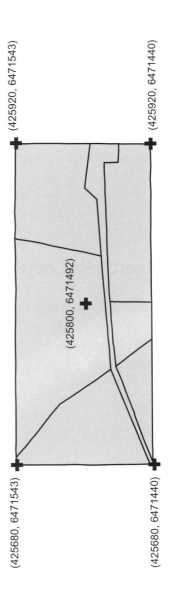

magnified when looking at the entire continent. Differences are apparent in length, width, and orientation. The projections displayed use global datum points, rather than datum points local to Australia. The centroid has been kept constant between the maps to allow comparison.

The geographic projection (Figure 5.4a) is not strictly a projection. It is a spherical reference guide that uses latitude and longitude coordinates.

Figure 5.4 3D Australia in 2D

a) Geographic projection

b) Sinusoidal projection

c) Equal-area cyclindrical projection

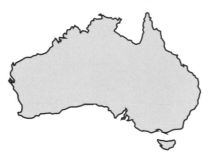

The sinusoidal projection maintains equal area although it introduces conformal distortion. It is free of distortion along the central meridian. The distortion is minimised when this projection is used for one land mass or at a global scale.

The equal-area cylindrical projection preserves correct area measurements. Shape, scale, and distance of features will be distorted towards the poles. This projection is generally most effective in an equatorial region.

Q4. Which projection appears to be more and more distorted towards the southern region of Australia?

CHAPTER SIX:
Data Output

Data output encompasses displaying results and producing an end product. It is this output product which will inspire an accolade, not the many hours devoted to data entry. It is not surprising that high quality data output devices and products are often towards the top of the budget list of priorities.

Data output

Regardless of how the data is to be used or manipulated in the GIS, there will be a call for data output at some stage. This may be a display on the monitor, a digital file, a map, a graph, a statistical table, a number, or a report.

Data output methods

As with data input, there are many methods and mechanisms for producing data output (Figure 6.1). The most common are listed below:

- *Drawing on the screen*: This is the cheapest and most easily updated output product. In fact, it is common now to take data and software on a computer to an agency or an employer and show the result to the client on the screen as a primary output product. The client can request alterations, which may be possible to make on the spot, before the map even touches paper as hard-copy output.
- *Printing or plotting onto paper*: The hard-copy map is still a product that makes the use and expense of GIS viable to many. A printed map, graph, table, or list is a hard-copy product that may represent much GIS modelling and analysis, or simply data entry. Knowledge of cartography is desirable for producing high quality displays.

- *Exporting data into a file*: This involves taking data, still in its digital format, and separating it from the GIS software. This is often done to transfer data to another computer software package, on to another computing system, or simply to store (back-up) GIS work. The file may be put onto a computer tape, a disk, electronically mailed, or stored on the computer hard drive. It is still the norm in many agencies and departments to use GIS for analysis, then export the resulting data to another software package designed expressly for creating graphs, maps, or reports.

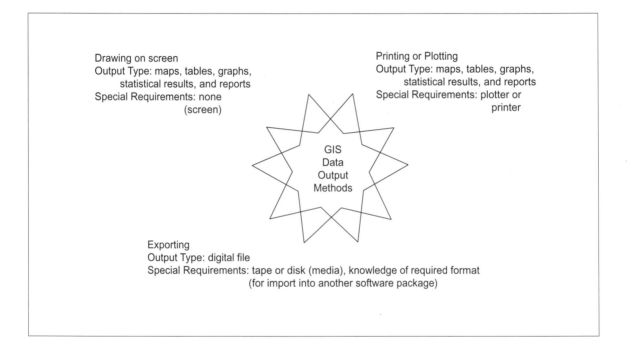

Drawing on screen
Output Type: maps, tables, graphs,
          statistical results, and reports
Special Requirements: none
          (screen)

Printing or Plotting
Output Type: maps, tables, graphs,
          statistical results, and reports
Special Requirements: plotter or
          printer

GIS
Data
Output
Methods

Exporting
Output Type: digital file
Special Requirements: tape or disk (media), knowledge of required format
          (for import into another software package)

Figure 6.1 The three main data output methods

Issues of data output

The main issues to be considered prior to data output include a clear identification of the user or interpreter of the data, knowledge of the best format, budget constraints, and general cartographic and presentation considerations (Figure 6.2).

If the data output is in the form of a display, either on screen or paper, consideration needs to be made of the people who will view and visually interpret the product. An effective species distribution

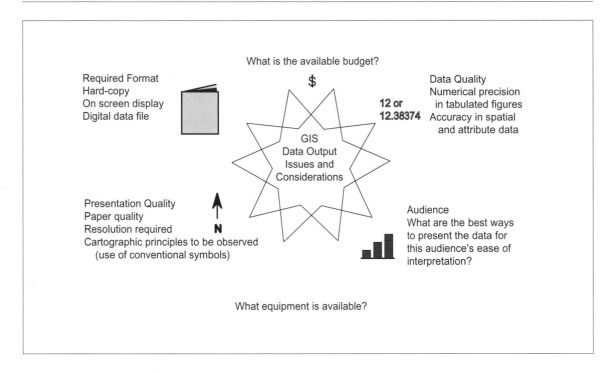

Figure 6.2 Issues and considerations concerning data output

map targeted at a scientific community with knowledge of the species may differ from a map of the same information produced for the general public.

Choosing the best output format may mean selecting to display a result as a graph or table, as a colour or black-and-white map, as a file able to be read by a PC or a different **computing platform**. Again, this will depend largely on the audience or the available hardware and software of the end user.

As with many aspects of GIS, there are various devices and accessories that can be used to produce output. A printer may be a cheap black and white line printer, colour pen plotter, or a laser printer. Paper types can vary from ordinary sized A4 to poster sized A0 and from flat matt paper to glossy photographic media. Export devices can range from floppy disk drives up to a complex system of external drives, and storage media can range from floppy disks to **compact disk**s.

With such a variety of choices, an appropriate budget should be set for data output. The user must keep in mind, however, that the success of a GIS is often gauged by the data output quality.

Finally, when producing maps in particular, good cartographic practices should be exercised. For example, a map should have a complete set of annotations and be constructed in a manner that will ease visual interpretation. In order, therefore, to become a proficient GIS user it is essential to have a basic understanding of key cartographic concepts.

Data output: Comments from the workplace

Without a doubt, the most popular use of GIS, apart from querying, is map making. This is so despite the advanced analytical and statistical toolboxes supplied in many GIS packages. Data output issues discussed and addressed by management instinctively tend to be related to a traditional map product. This is not surprising—geographical information is considered by many as synonymous with maps. The following comments were typical of the user responses.

> People do want maps and the more colourful the better. And I don't think this will change in the future.
>
> *Werner Runge, GIS Project Officer and Consultant*

> Most managers or directors place a great emphasis on maps. 'How else can you see results?' they often say.
>
> *Piers Higgs, GIS Consultant*

Among GIS users who enjoy data analysis, there is a tendency to use GIS as a workhorse and to plot the results using alternative software for presentation purposes. For example, a landscape architect may put their map into a visualisation package in order to produce a visually pleasing, realistic looking result. A statistician may prefer a graph of a particular scientific style unavailable in GIS.

> I have recently used the tools available with a GIS for creating graphs, but found the tools rather limiting. I feel that other specialised software packages would perform this task better.
>
> *Narah Stuart, GIS Modeller*

The tools developed within GIS have been 'learned' from a number of fields and often are, necessarily, simplified examples of sophisticated tools found elsewhere.

Exercises: Getting the Jarrahlea data out of your system

Figure 6.3 shows data output from the Jarrahlea data set. Specifically, there are examples of a map product, a chart, a report, a table, and a statistical result. Imagine you have been asked to conduct the GIS tasks listed below. Suggest which of the final product formats would meet the needs of the task and list any output devices you may need. Note that the data displayed in Figure 6.3 is not the data you would require for the tasks. These are representations of possible output formats.

Q1. You have been requested to show the user the variability of soil types, spatially, over Jarrahlea.

Q2. The manager of your department has asked for an indication of the relative areas occupied by each soil type on Jarrahlea.

Q3. The owner of Jarrahlea would like to know whether olive farming would be a feasible land-use for his property. You should supply him with any relevant information for a feasibility analysis.

Figure 6.3 The many faces of data output for Jarrahlea

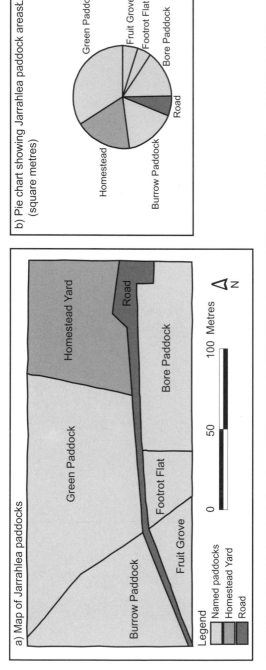

a) Map of Jarrahlea paddocks

b) Pie chart showing Jarrahlea paddock areas‚ (square metres)

c) Report on Jarrahlea property

Jarrahlea: Potential for Soil Erosion

d) Tabular Jarrahlea paddock data

Data Element	Name	Area (m²)	Perimeter (m)
Polygon	Green Paddock	8289	1339
Polygon	Burrow Paddock	4132	413
Polygon	Bore Paddock	3719	301
Polygon	Footrot Flats	1118	140
Polygon	Fruit Grove	1218	202
Polygon	Homestead	4519	282
Polygon	Road	1519	530

e) Statistical result showing‚ the average area (square‚ metres) of paddocks on‚ the Jarrahlea property

3502

CHAPTER SEVEN:
Data Management

Data will not look after itself. Without database maintenance, employing a **database management system (DBMS)** and continually updating data dictionaries and metadata, GIS stands a good chance of disarray and failure.

Most GIS software supply the ability to control data and databases using a DBMS. Having these controls over the computational organisation of the data usually occurs without the need for the user to know anything about database storage or DBMS techniques.

The GIS data management processes that users must be aware of are the creation and updating of data dictionaries and metadata. Following guidelines in these two areas will allow greater data reliability and ease of access for the larger GIS community.

Data organisation within GIS

By now the reader will realise that it is possible to have an enormous amount of data within a GIS. There are levels of organisation of this data. The purpose of this organisation is to ensure stability, consistency, and reliability, and to maintain the data in a way that will allow greatest flexibility in GIS data manipulation, analysis, and modelling.

One organisational term, which has already been used in this text, is 'layer'. A digital layer will contain one piece or type of data and generally only one element of data structure.

In vector GIS a layer may contain points, lines, or polygons (Figure 7.1). Each layer represents an entity (something we wish to describe), or a form of an entity, distributed across space. For example, the following layers may be derived from a topographic map:

- Point layers: wells, spot heights, farm buildings
- Line layers: contours, roads, rivers, power lines
- Polygon layers: lakes, Local Government Areas, land use.

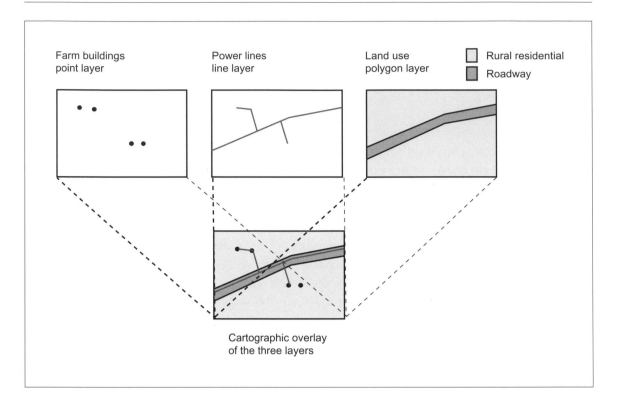

Farm buildings
point layer

Power lines
line layer

Land use
polygon layer

☐ Rural residential
■ Roadway

Cartographic overlay
of the three layers

Figure 7.1 Vector GIS layers

Notice that one grouped entity, water-related features, has been broken into three separate elements (wells, rivers, and lakes) based on the different inherent data types (points, lines, and polygons) at one scale.

It is advisable to keep data in the simplest form when creating a database. It is a quick and easy GIS process to merge layers into more complex layers; however, it is time consuming to divide a layer into two more fundamental layers. The aim in building the initial database is to keep each layer as a unique building block for processing, analysis and modelling. Working with small building blocks ensures flexibility in data use.

Raster GIS works on much the same principle. A **grid layer** (or grid cell layer, or cell layer) is a unique layer of data (Figure 7.2). Instead of points, lines, and polygons, the grid layer is a collection of equally sized and shaped cells. Each cell in a grid holds data, represented as numbers, relating to the spatial entity being represented. Grid data should be maintained at the most appropriate cell size to achieve the highest possible resolution. It is easy to generalise to a larger cell size in raster GIS, if required.

Each of the layers listed in the section describing vector layers could be represented as a grid, regardless of the original type (point, line, or polygon). The success of the representation, analysis, and display of the grid cell layers will depend on the selection of an appropriate cell size.

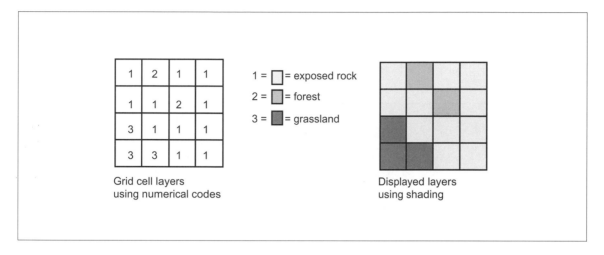

Grid cell layers using numerical codes

1 = □ = exposed rock
2 = ▨ = forest
3 = ▦ = grassland

Displayed layers using shading

Databases

Figure 7.2 Grid cell layers

As the amount of data grows, the need to organise data becomes apparent. For example, a GIS user may well find they have elevation contour and population density data for Sydney and Perth, which are spatially located approximately 4000 kilometres apart. The layers are called 'contours' and 'popdensity'. How will it be possible to differentiate between the two layers? The answer is to use separate databases (Figure 7.3).

Databases are collections of related data. The relationship is usually based on location (for example, a Perth database); however, they could be based on the data structure and type (such as a vector points database), or a common underlying purpose (for example, the urban planning database).

Think of a database as either a folder or a map cabinet within which many different layers and grid cell layers may be stored (Figure

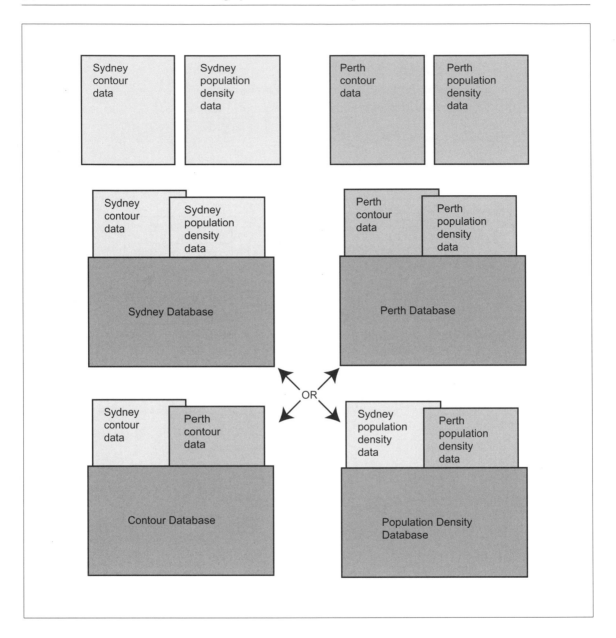

Figure 7.3 Databases

7.4). The folder or cabinet will contain all the necessary spatial and attribute data, along with topology, stored as layers.

The Jarrahlea database represents a very simple **relational database**. Relational databases consist of two-dimensional tables (termed relations). Columns in an attribute relation are attribute headings,

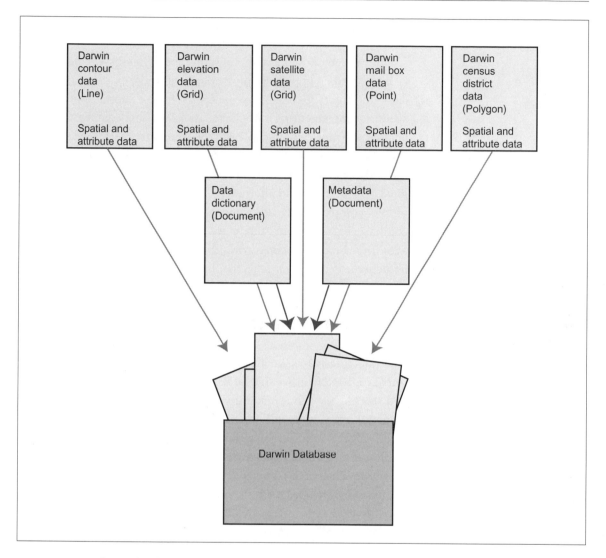

Figure 7.4 Databases contents

rows are records, and cells are attribute values. Spatial data may be stored similarly in one of these tables. The spatial and attribute tables are 'related' through a common attribute, such as a GIS-assigned record number, or ID number, as mentioned in the exercise in Chapter Four.

There are a number of other types of database structures. Examples include **hierarchical** structures, with a parent–child, or one–many relationship, and **object-oriented** structures, which demonstrate object inheritance. Further description of these different database structures is beyond the scope of this introductory text. The reader is directed to the references for further information.

The choice of the most appropriate type of database structure will be best based on knowledge of the data, interrelationships within the data, and the types offered by the GIS software chosen. Many larger agencies implementing GIS have an established database in specialised software, which can easily be integrated with GIS software. Using specialised database software may allow the agency to exploit the additional abilities on offer in these packages, or data may simply have been stored in an alternative system prior to the establishment of the company GIS department. Maintaining this independent database ensures consistency in database principles over time, avoiding the cost of converting the data into a GIS database, and allowing users easy access to the data for many purposes other than GIS.

Database Management Systems (DBMS)

The database management system (DBMS) is the part of the GIS software that organises the database. This oversees all input, output and storage, as well as guiding access to the data. A good database management system will unobtrusively play a vital role in the GIS. The DBMS will be unobtrusive as the user will not be aware of the mechanisms used to manage the data, only how to affect and alter DBMS management parameters. The DBMS will play a vital role in ensuring data integrity, consistency, and security. The DBMS should also allow for concurrent use of data, if this is necessary, and ensure efficiency in all aspects of database management.

The DBMS will reflect the style of the database. A relational database will require a relational database management system. It follows that a DBMS may be integrated into GIS software or may be independent of software.

Data dictionaries and metadata

Historical and present knowledge of the data is crucial to all GIS work. The scale of the data, the key to data codes, or the date of data capture can be vitally important in any GIS process from producing a display through to modelling. This information should be documented in a **data dictionary** and be available as metadata. Data dictionaries and metadata will guide the user through selecting appropriate data to determining appropriate processing and analysis techniques. The only situation worse than using imprecise and inaccurate data is to be using data without a known past.

Data dictionaries are documents that contain a key to data in a database. Generally a table is used to list explanations of codes and abbreviations, which constitute the data dictionary. For example, a column of code numbers (1, 2, 3 . . .) used in a grid cell layer may be given beside a column of land-use category names (industrial, residential, rural . . .) which are not stored in the GIS database. It is relevant to point out that this information could be compiled into a relational table so that if the names are needed in a GIS process involving a relational database and a relational DBMS they can be easily accessed.

Metadata refers to data about the data. A metadata document contains a description of relevant details, such as the date of creation, the custodian, the original (non-digital) source, the currency, the status, how to access the data, data quality, and even cost for potential purchasers.

Often metadata will contain a data dictionary, the two being combined into one digital or hard-copy document.

Data management: Comments from the workplace

The surveys of GIS users indicated that data management and maintaining data dictionaries and metadata are crucial, although their usefulness has been underestimated in Australian situations in the past.

The Australian GIS community has a great deal of data on computers 'here, there, and everywhere'. Perhaps it is a function of the remoteness of some centres, such as Perth, Darwin, and Hobart, which acted as a barrier to organised standards in databases, DBMS, data dictionaries, and metadata. More likely is the proposition that Australians, like global GIS communities, participated in frantic data capture during the establishment of GIS departments from the

1970s to the early 1990s, when any digital data was seen as good digital data and an asset. The realisation that 'good' or 'appropriate' digital data is a subjective tag that must be determined by the user in the light of their application has led to the introduction of metadata and data dictionary guidelines. The Australian versions are based largely on overseas guidelines and standards. While metadata guidelines are not enforced or policed in Australia, there is a general feeling that cooperation will lead to a responsible GIS community and to improved access to data.

Below are comments from several GIS users. The first reinforces the need for a good DBMS and general management techniques; the second and third comments discuss the need for data dictionaries and metadata in order to support GIS. It is obvious that a lack of important documentation and appropriate data-handling strategies have caused these people tremendous difficulties when using data that was established prior to their employment. Australian GIS users have strong views about databases, database management systems and appropriate data documentation.

> After working in an environment where metadata was not considered because it was thought it was a peripheral and not necessary, I'll say it's vital. In workplaces where [information technology] investment is a bit low, across platform linkages to databases that reside on UNIX machines are neglected. This means that for someone to work on a Windows machine with the same data it needs to be exported as text. This results with several out of date sets of the same database floating around the company.
>
> *Nick Middleton, GIS Consultant*

> Data management is an essential part of any GIS and, contrary to what most management thinks, is not just storing data on disk. Too often people ask their tech support for data without understanding that these people have data dictionaries in their heads! And when the tech support leave, there go your data dictionaries. Metadata, data dictionaries, update strategies and corporate GIS strategies are all part of a GIS. They should be treated as important as the data itself—without them, the data is basically useless!
>
> *Piers Higgs, GIS Consultant*

High quality metadata and data dictionaries are as important as high quality data. I find the ANZLIC [Australia New Zealand Land Information Council] metadata guidelines very thorough. Although they may initially appear a bit too detailed, in the long run it does serve you well. My personal experience has been with a dataset that I generated from a relatively large number of sources. Keeping detailed metadata has saved me considerable time and embarrassment when people have needed some information on the source of the data. Detailed data dictionaries are also vitally important for other people to interpret your data, especially when using codes to assign attributes. Maintaining an adequate data dictionary does take a bit of discipline!

Narah Stuart, GIS Modeller

The lesson to learn is that a responsible GIS user will obtain and update data dictionaries and metadata for all the data they use.

Exercises: About the data about the data

Let us now look at gleaning information from metadata and data dictionaries. Figure 7.5 has a map, a data dictionary, and an extract of metadata. Study this figure in order to answer the following questions.

Q1. Identify which part of Figure 7.5 (part b or c) constitutes a data dictionary, and which constitutes a metadata extract.

Q2. Why do you think the data is not due for completion ('Date Ending:' in Figure 7.5c)?

Figure 7.5 About the data about the data

a) Map of paddocks showing data related to exposed soils

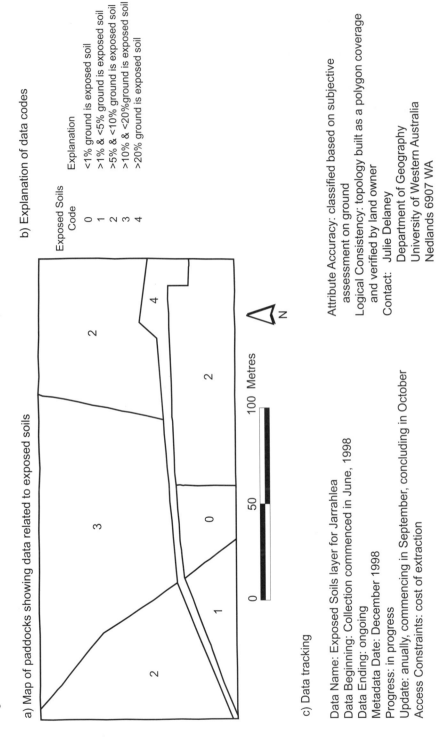

b) Explanation of data codes

Exposed Soils
Code   Explanation
0   <1% ground is exposed soil
1   >1% & <5% ground is exposed soil
2   >5% & <10% ground is exposed soil
3   >10% & <20%ground is exposed soil
4   >20% ground is exposed soil

c) Data tracking

Data Name: Exposed Soils layer for Jarrahlea
Data Beginning: Collection commenced in June, 1998
Data Ending: ongoing
Metadata Date: December 1998
Progress: in progress
Update: anually, commencing in September, concluding in October
Access Constraints: cost of extraction

Attribute Accuracy: classified based on subjective
    assessment on ground
Logical Consistency: topology built as a polygon coverage
    and verified by land owner
Contact: Julie Delaney
           Department of Geography
           University of Western Australia
           Nedlands 6907 WA

CHAPTER EIGHT:
Elementary GIS Tools

This chapter describes the tools that will equip the user with basic GIS functionality. These elementary tools represent stepping stones into more complex analysis and querying. Elementary tools are ideal for exploratory data analysis, i.e. having an initial inspection of the data.

The basic toolbox

The data is captured, the editing is complete, topology has been built, registration and projection are established, and the user is ready to commence employing the GIS tools.

The tools discussed initially, in this chapter, are fundamental tools that should be available in all GIS software packages. Some of the tools presented are also available in software packages other than GIS.

Listing

The first task the GIS user may wish to perform is to check that the tabular data appears to be correct and consistent. This may involve listing the spatial coordinates, the associated attributes, or both (Figure 8.1). A list function is usually supplied with the DBMS. The DBMS facilitates listings of vector and grid layers available, attribute definitions, and attribute values by record number, or grid cell location. A spreadsheet may offer very similar functionality and an output of a similar appearance.

Displaying

Displaying the data spatially is another primary use for the GIS (Figure 8.2). The user can draw data in a graphic display and control

Give a listing of ...

a) Available layers	b) Attribute definitions for the wells layer			c) Attribute values for the wells layer		
Polygon						
land use	Name	Type	Length	Id	Depth	Use
soil units				1	10.34	Water sampling
Line	Id	Number	8	2	25.83	Drinking water
tracks	Depth	Number	8	3	13.48	Irrigation
pipe lines	Use	Text	20	4	23.37	Water sampling
Point				5	34.97	Irrigation
wells						
soil sample sites						
Grids						
elevation						
temperature						

Figure 8.1 Listing

how the data will appear. The links between the spatial and attribute data allow the user to create an appropriate shading scheme for display. The scheme may be based on values of an attribute or some other form of symbolisation. For example, the user can create a map of soil units, shaded using the soil unit pH level. The shades may vary from red (low pH) to blue (high pH). Furthermore, the user can also put the number related to pH level on the map, or create a key.

Although display is a form of data output, this may also be used as a tool of exploratory data analysis, aiding visual interpretation. Display, therefore, need not be limited to maps. Graphs and lists, for example, may be termed displays.

Figure 8.2 Displaying

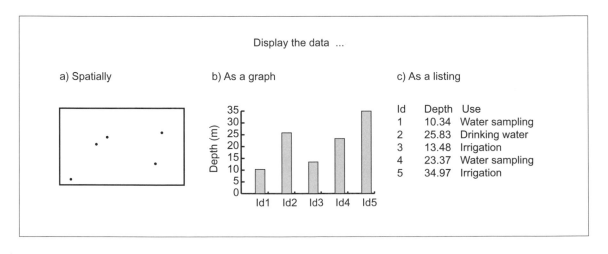

Display the data ...

a) Spatially   b) As a graph   c) As a listing

Id	Depth	Use
1	10.34	Water sampling
2	25.83	Drinking water
3	13.48	Irrigation
4	23.37	Water sampling
5	34.97	Irrigation

Results similar to a GIS display can be obtained using computer cartography, graphing packages and spreadsheet software.

Querying

Listing and displaying data may lead to the desire to query the data (Figure 8.3). The user may issue a query based on attribute values and spatially identify a solution. An example of such a query may be to query the layer of potential foster homes to show the GIS user the spatial locations of only those homes with a certain religion and family size.

Figure 8.3 Querying

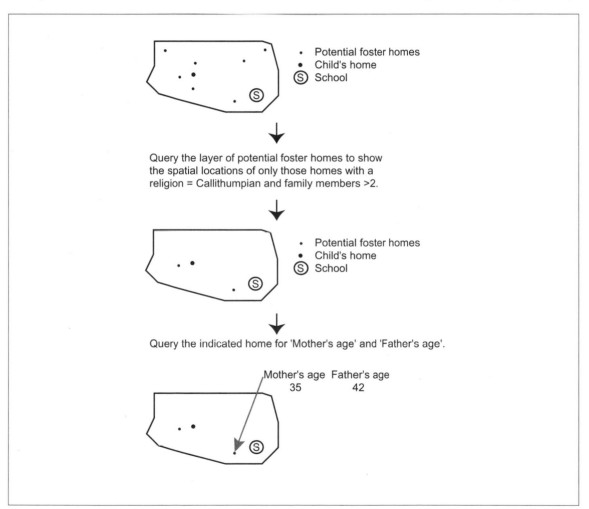

The query would be worded something like this: 'Select all potential foster homes with religion equal to Callithumpian and family members greater than 2.'

The user may view the spatial distribution of the result of this query, or decide to list the selected records.

Alternatively, the user may query an attribute value at a spatial location by pointing to the location with the mouse. The term 'display querying' can also be used to describe this 'point and click' technique. Perhaps, in the above example, the user is interested in a potential foster home simply because it is spatially closer to the child's school than other homes. By pointing and clicking on the home of interest, all the relevant attribute data for that home can be displayed.

These query methods can be used for both raster and vector data, although the manner in which the query is constructed will differ.

Each GIS software package will have a set of tools to allow querying. Some software packages will allow querying of lists and spatial displays simultaneously.

Reclassifying

Querying and other exploratory data analysis may suggest that a new data classification would be more appropriate or would illustrate a phenomenon more clearly (Figure 8.4). It may be that the user is planning the location of a new picnic area in a forest. Slope appears to be the major limiting factor, as most of the landscape is highly dissected. The user can add a new attribute (column of data) and reclassify the current data using a scheme, such as: 'If the site is less than 5% in slope, let the new attribute (potential picnic site) be equal to yes. Otherwise let the new attribute (potential picnic site) be equal to no.'

Reclassification may be based on numeric or text attributes, or a mixture of both. It is usual to phrase the query like a mathematical expression, regardless of the type of attribute.

Similar reclassification functionalities are available in most spreadsheet packages, although a GIS is essential in order to see the spatial result.

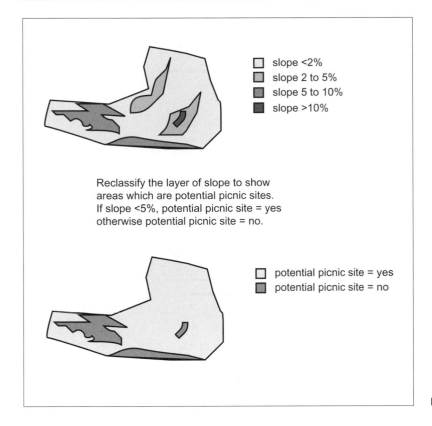

slope <2%
slope 2 to 5%
slope 5 to 10%
slope >10%

Reclassify the layer of slope to show
areas which are potential picnic sites.
If slope <5%, potential picnic site = yes
otherwise potential picnic site = no.

potential picnic site = yes
potential picnic site = no

Figure 8.4 Reclassifying

Measuring

Measuring is a form of querying. Measuring generally involves a
selection and a calculation (Figure 8.5). An example would be:
'Travel down Geographical Place, turn right into Information
Avenue, and then left into System Street. At the end of System
Street, measure how far we have travelled.' The selection is the list
of streets to be travelled, and the calculation is the addition of the
lengths of the selected streets.

Measuring is not necessarily a text-based query requiring a statement. Some GIS software packages would allow the user to click on
the street on a display (selection), and issue a measurement command or click a measurement button (calculation), to arrive at the
same measured result. A similar process can be undertaken for area,
perimeter, and any other measurable variable.

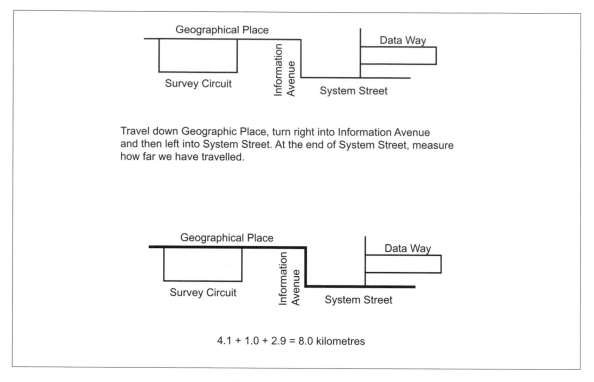

Travel down Geographic Place, turn right into Information Avenue and then left into System Street. At the end of System Street, measure how far we have travelled.

4.1 + 1.0 + 2.9 = 8.0 kilometres

Figure 8.5 Measuring

An alternative way of measuring is to use the spatial display and geographical information only. Most GIS software packages offer a ruler tool, which will allow you to click anywhere on the display (a virtual node) and continue clicking (virtual vertices), drawing a complex line. On completion the line is removed from memory and the length displayed. The measurement may be in screen units or map units or, for example, metres if using MGA94. Other popular measurement tools include a user-defined rectangle (measuring area or perimeter) and a user-defined circle (measuring area, circumference, or radius).

Measurements made in raster GIS are given as a number of cells. Converting this to area involves multiplying the number of cells by the area of a cell.

Reporting

The final elementary tool we will discuss is the reporting functionality of the GIS (Figure 8.6). This is a statistical tool that calculates

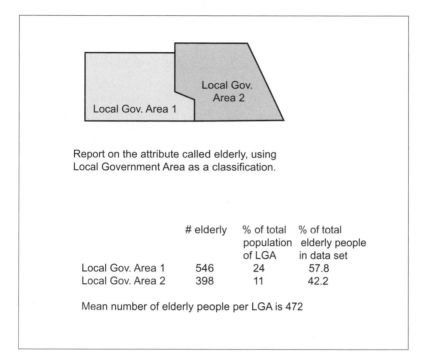

Report on the attribute called elderly, using
Local Government Area as a classification.

	# elderly	% of total population of LGA	% of total elderly people in data set
Local Gov. Area 1	546	24	57.8
Local Gov. Area 2	398	11	42.2

Mean number of elderly people per LGA is 472

Figure 8.6 Reporting

summary statistics for a layer, or for a subset of data from a layer. An example would be: 'Report on the attribute called elderly, using Local Government Areas as a classification.'

Common statistics given in a report include the mean, standard deviation, mode, minimum value, maximum value, and the range.

This tool may be an important starting point for a more detailed analysis using the tools which are discussed in Chapters Nine and Ten.

Elementary GIS tools: Comments from the workplace

The tools described in this chapter are undoubtedly simple. Many similar tools may be obtained in other software packages, which are often more accessible and more suited to a specific form of viewing or querying data.

> I think data reporting tools are lacking in the [GIS] software packages
> I've used. When publishing reports, companies like to see textural data

nicely formatted. This usually results in exporting data out of the GIS database and into some (other software) package.

Nick Middleton, GIS Consultant

For spatially based elementary tools, such as querying data using a spatial or geographical reference, and visualisation of a query, there is no comparable alternative to GIS. Wide acceptance of these levels of GIS capabilities gives confidence to GIS users.

Displaying information is the most basic GIS tool I have used the most often. After that, almost all analysis I have conducted has involved reclassification of some sort, such as aggregating detailed data into broader categories. I have also used querying a lot as a part of analysis, not as a simple query on a dataset.

Narah Stuart, GIS Modeller

With experience, Australian GIS users have become aware that these basic tools can be the start of a more complicated analysis.

Exercises: Listing, displaying, querying, reclassifying, measuring, and reporting on Jarrahlea

Figure 8.7 contains some examples of the elementary tools we have been discussing in Chapter Eight.

Q1. Which elementary GIS tools can you identify in Figure 8.7? For each tool, give a short description of how it is being used.

Q2. Which elementary GIS tools can you identify in Figure 8.8? For each tool, give a short description of how it is being used.

Figure 8.7 Elementary GIS tools and Jarrahlea data

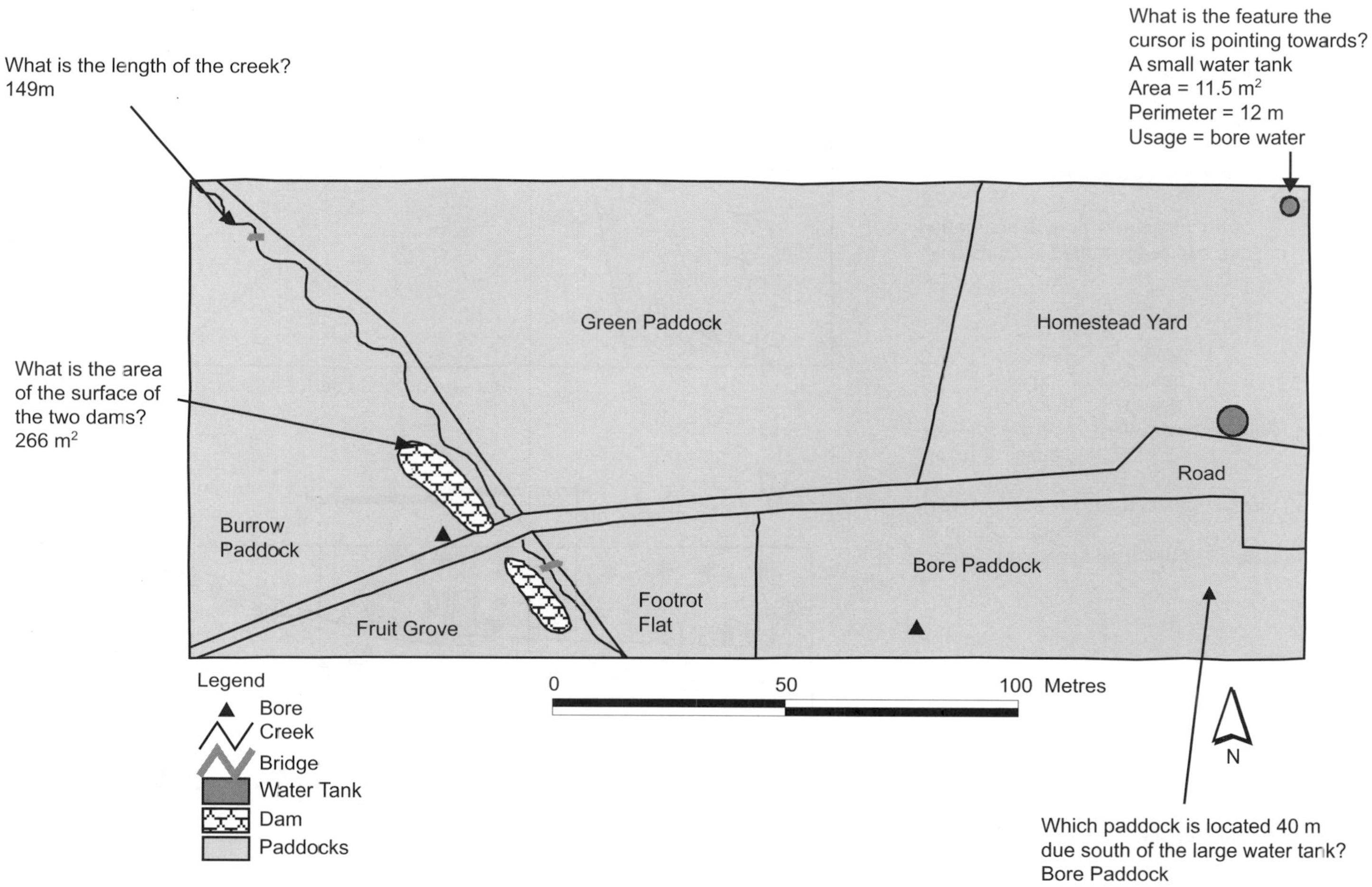

What is the feature the cursor is pointing towards?
A small water tank
Area = 11.5 m²
Perimeter = 12 m
Usage = bore water

Homestead Yard

Road

Bore Paddock

Green Paddock

Footrot Flat

Fruit Grove

Burrow Paddock

0   50   100 Metres

Which paddock is located 40 m due south of the large water tank?
Bore Paddock

N

What is the length of the creek?
149m

What is the area of the surface of the two dams?
266 m²

Legend
Bore
Creek
Bridge
Water Tank
Dam
Paddocks

Figure 8.8 More elementary GIS tools and Jarrahlea data

a) Base map of marri and fruit trees

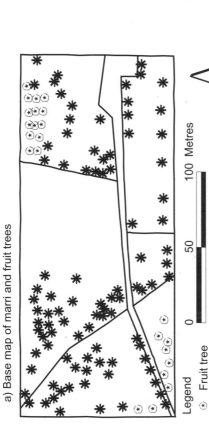

Legend

⊛ Fruit tree

✱ Marri tree

▭ Paddocks

0   50   100 Metres

N

Can we alter the marri and fruit tree display to show the spatial distribution of young, moderate and old aged trees?

Result is displayed in Figure b)

Supply the details pertaining to marri trees.
Available attributes:

Tree-Id	Age	Crown cover	Height (m)

Records listing

Tree-Id	Age	Crown cover	Height (m)
1	Young	dense	6
2	Old	sparse	32
3	Young	sparse	2

Give summary statistics relevant to the vegetation in the Homestead Yard

33 of 33 (100%) trees are old trees
14 of 33 (42%) trees are fruit trees
19 of 33 (58%) trees are marri trees
14 of 14 (100%) fruit trees are old trees
19 of 19 (100%) marri trees are old trees

b) Reclassified map showing tree age

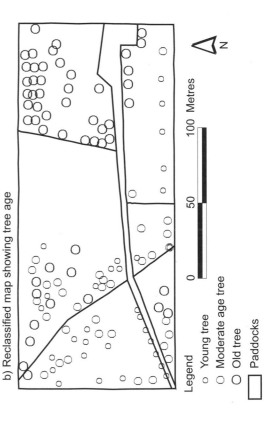

Legend

∘ Young tree

○ Moderate age tree

◯ Old tree

▭ Paddocks

0   50   100 Metres

N

CHAPTER NINE:
Vector Geoprocessing Tools

Vector geoprocessing tools manipulate data based on a user definition relating to attribute data or spatial extent. The result of using a
geoprocessing tool is a new layer. The new layer does not generally
contain new data, rather a subset or reorganisation of the original
layer's data. Overlay, however, combines two layers into a new third
layer, which will contain spatial and attribute data from both original layers.

These tools are ideal for creating a secondary database, or generalised layers for a specific purpose.

Vector geoprocessing tools

Geoprocessing tools form part of the GIS toolbox and are specifically
designed for the manipulation and analysis of spatial and attribute
data. Geoprocessing tools alter pre-existing data to create new
(derived) layers and attribute tables (Figure 9.1). The creation of a
new layer distinguishes these tools from the elementary tools discussed in the previous chapter. The original data is always retained,
unaltered, in the original layer or layers.

Most GIS software packages will supply all the tools discussed in
this chapter, although the name allocated to the tool may differ from
package to package. Some packages offer geoprocessing tools not
mentioned here. These 'extra' tools are generally designed for a specific purpose, and are a variation on one or more of the tools
described in this chapter.

Geoprocessing tools are available for use with vector or raster GIS
data. This chapter describes vector geoprocessing tools only. Chapter
Ten will introduce raster geoprocessing tools.

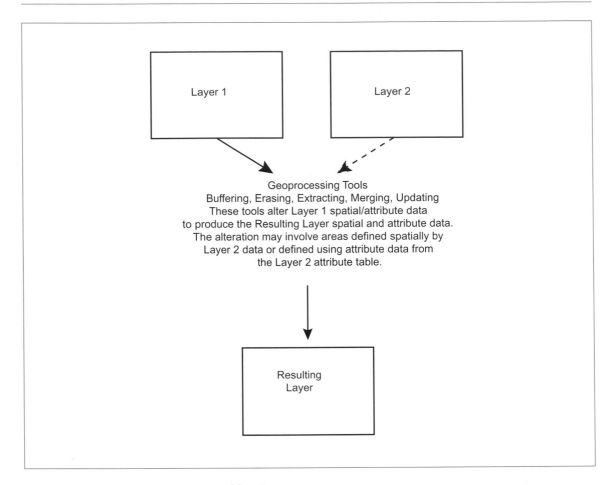

Figure 9.1 Geoprocessing tools

Buffering

Buffering is a process that creates an area (polygon) around a feature (Figure 9.2). The result of any buffering process is a new polygon layer.

To implement a buffer process, the GIS user designates a distance, say 100 metres, and a layer, say rivers. In this example, all the land within 100 metres of any river will be contained in a new polygon layer. This may be used, for example, to designate a riparian vegetation protection zone.

It is possible to define a variable buffer distance based on an attribute. For example, a user may have a roads layer (containing main and minor roads) and decide to define buffer zones between

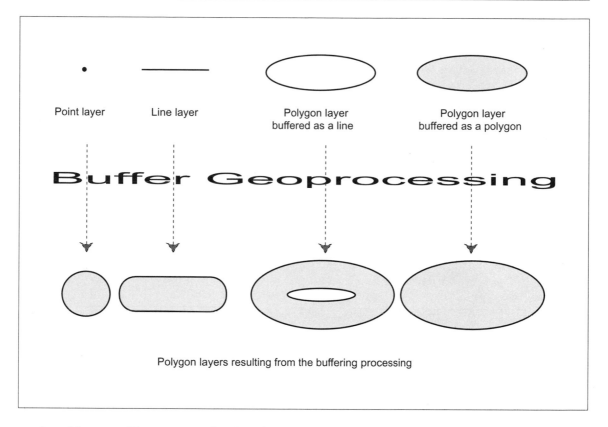

Point layer   Line layer   Polygon layer   Polygon layer
                                         buffered as a line   buffered as a polygon

Buffer Geoprocessing

Polygon layers resulting from the buffering processing

roads and housing. The user may decide to buffer main roads to a distance of 20 metres and minor roads to a distance of 10 metres.

Figure 9.2 Buffering

Erasing

Erasing, or deleting, data from a layer to create a new layer can be
implemented in one of two main ways (Figure 9.3).

The simplest erasing method involves just one layer. An example
may be as follows: a layer of National Park lands exists and a planner
is selecting possible sites for new tourist drives. The drives must not
pass through wilderness areas. Therefore, the GIS user will erase
these unsuitable areas prior to addressing further considerations. The
command is given to create a new layer with all wilderness areas
erased. An erase statement may read: 'Find all National Park polygons that have an attribute called wilderness (possible attribute
settings being either 'yes' or 'no') equal to yes. Erase these polygons

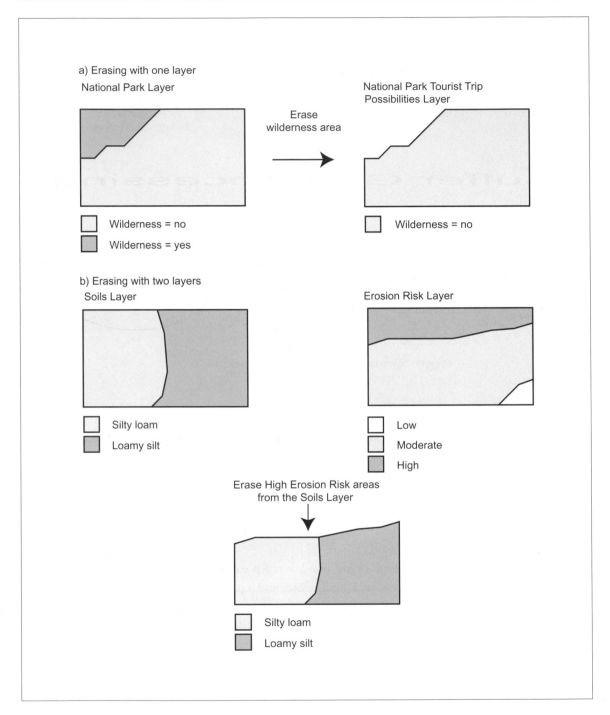

Figure 9.3 Erasing

to create the new layer called 'National Park Tourist Trip Possibilities'.'

The second method involves two pre-existing layers. The process deletes data from one layer, defined by data in the second layer, to create a third, new layer. For example, imagine that a database contains layers of soils and erosion risk zones. 'Possibly arable soils' need to be located for agricultural expansion zones. The project manager has specified that the 'possibly arable soils' cannot have a high erosion risk. The first layer contains the information of interest and the data required in the result (soils). The second layer contains the selection criteria for the erasing process (erosion risk). The GIS user knows that the area defined by any polygon in the erosion risk layer with risk equal to 'high' can be excluded from all future analysis. The erase process therefore removes any areas on the soils layer that are contained within polygons with a 'high' attribute label in the erosion risk layer. The resulting soil data are put into a new layer (Figure 9.3b).

It may help to envisage the polygons defined by the selection rules (erosion risk = 'high' in the above example) acting as 'cookie cutters' to clip out, or erase, their own spatial extent from another layer (soils) to create the third, new layer.

Extracting

If erasing methods act as cookie cutters to remove unwanted pieces of a layer (removing the cookie and keeping the dough for further work), the extracting methods do the reverse: maintaining only specified areas (keeping the cookie and discarding the dough). Extracting and erasing can be used to obtain identical results.

Extracting data from a layer to create a new layer can be implemented in one of two ways (Figure 9.4), similar to the erasing example given previously.

The first method involves identifying data to be extracted from an existing layer. If a planner was asked to make a map showing only the green space in a city, the extract geoprocessing tool may be used to extract these areas from the land-use layer and place them into a new layer (Figure 9.4a). An extraction statement may be formed as follows: 'Find all areas in the land-use layer that are green space. Extract

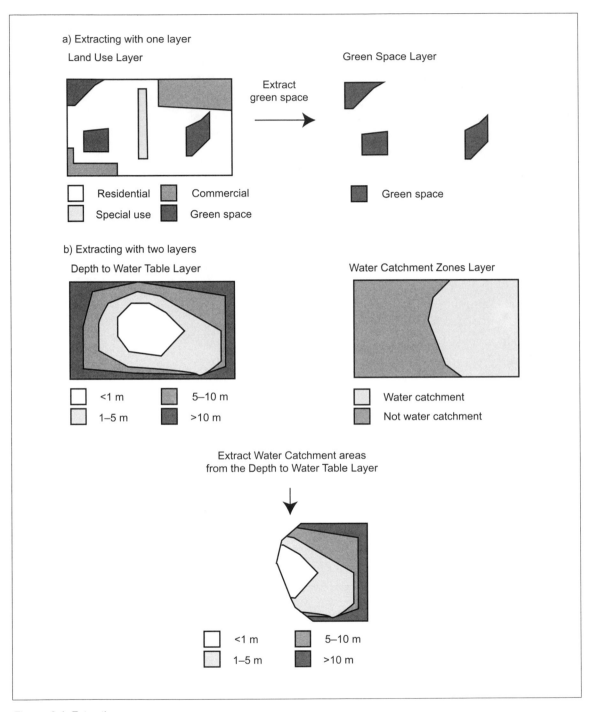

Figure 9.4 Extracting

their spatial and attribute data and place the copy in a new layer called 'green space'.'

The second method involves two layers, for example, depth to water table and water catchment zones (Figure 9.4b). Imagine that a water authority needs to calculate the average depth to water table in the catchment area feeding into a dam for a groundwater quality study. One layer, depth to water table, contains the data of interest. The second layer, catchment zones, defines the area of interest. The extracting process therefore isolates any areas in the water table layer that are contained in the water catchment zones and the result is put into a new layer. The user can then employ an elementary GIS tool to find the average depth to water table.

Merging

In **merging**, polygon boundaries are dissolved in order to merge adjacent polygons (Figure 9.5). The decision to merge two polygons or to keep them unique is based on their attribute values or their ability to satisfy set criteria.

Figure 9.5 Merging

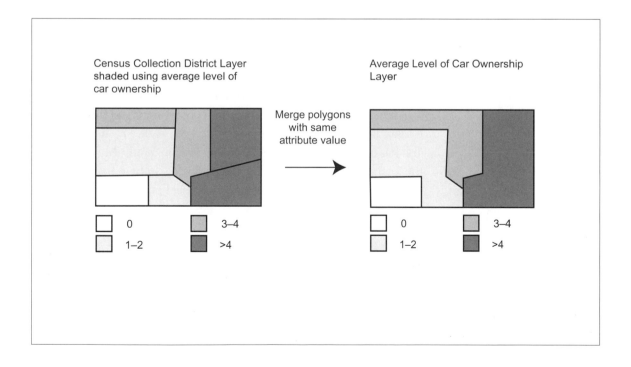

A prime example uses census data. In any one city there are many census collection districts. A collection district layer shows the spatial boundaries of each district. This layer may have a large attribute table containing all the census data. A transport analyst may wish to have a separate layer containing only attribute data relevant to the average level of car ownership. This is an attribute of particular interest whereas the other attributes may be irrelevant to the transport analyst. Creating a unique layer would lower the time spent manipulating and processing data. Quite often the case will be that two bordering polygons (census collection districts) have the same average level of car ownership. This renders the border between the polygons essentially meaningless to the transport analyst. A merge operation will remove the common boundaries between census collection districts that have the same attribute value, thereby simplifying the resulting display.

A common use of merging is the removal of sliver polygons. This is generally done by conducting a search for polygons of very small area. The largest side of the small polygon is dissolved, merging the sliver polygon with the neighbouring polygon.

Updating

Updating a layer is a common process in dynamic data sets. This allows spatial and/or attribute data associated with a section of the layer to be updated based on new or more relevant data (Figure 9.6).

An example may be that a planning department has the most up-to-date land-use layer available (updated weekly). The local council also has a land-use layer, which they update on a monthly basis using the planning department's data. The local council has attributes attached to their land-use map, such as local council recommendations for future development, which are entered after obtaining the data from the planning department. Instead of erasing the entire layer and replacing it with a new layer each month, which would mean re-entering all the attribute data unique to the local council, the GIS users can employ an update function.

The process works in the following way. The planning department sends the local council a layer called, say, 'changed', which contains only the polygons of land parcels that have changed in land use in

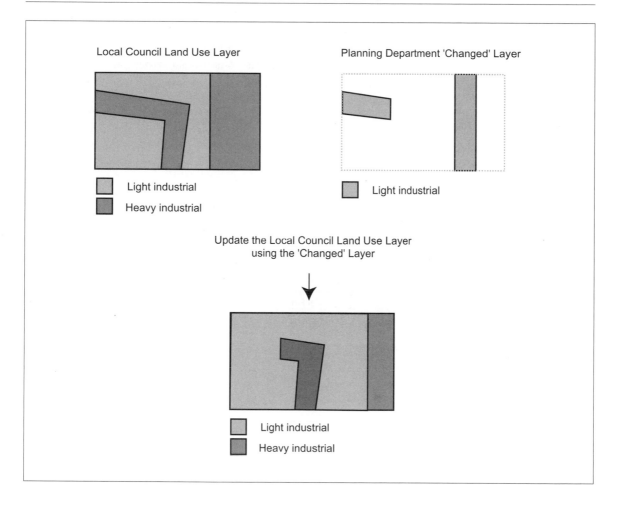

the past month (they may have used an extract operation to create this layer). The local council updates their land-use data using this 'changed' layer. Their GIS finds the spatial location of any polygons in 'changed' in the local council land-use layer, erases the current data, and inserts the new data.

This may be envisaged as a spatial and attribute cut and paste operation.

Vector overlay

Overlay is perhaps the most commonly used and best-understood operation in GIS application. It is, as the name suggests, laying, or

Figure 9.6 Updating

superimposing, one data layer directly over another (Figure 9.7) for the purpose of exploring relationships and associations in the data. If no new data are created, this constitutes cartographic overlay. If a new layer results from merging two or more data layers, this is true

Figure 9.7 Vector overlay

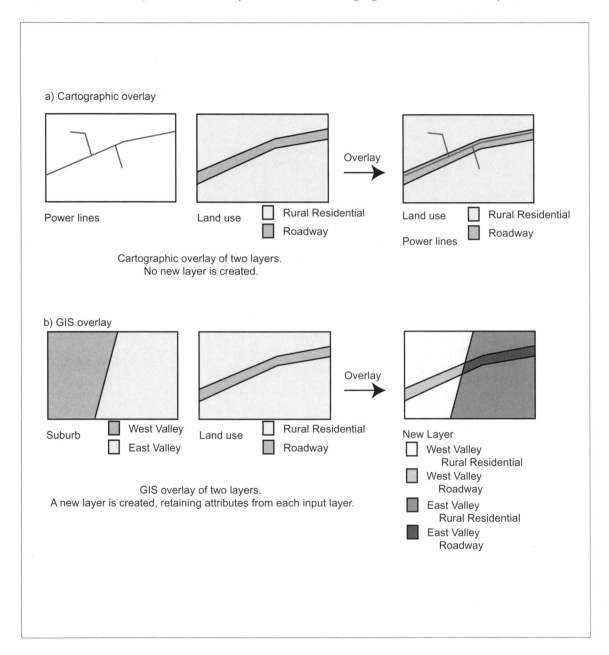

a) Cartographic overlay

Power lines

Land use ▫ Rural Residential ▪ Roadway

Overlay

Land use ▫ Rural Residential ▪ Roadway

Power lines

Cartographic overlay of two layers.
No new layer is created.

b) GIS overlay

Suburb ▪ West Valley ▫ East Valley

Land use ▫ Rural Residential ▪ Roadway

Overlay

New Layer
▫ West Valley
   Rural Residential
▪ West Valley
   Roadway
▪ East Valley
   Rural Residential
▪ East Valley
   Roadway

GIS overlay of two layers.
A new layer is created, retaining attributes from each input layer.

GIS overlay. The overlay procedure differs between vector and raster data.

Vector overlay is a process commonly used by cartographers, landscape architects, geographers, and many other users of spatial data. Traditionally it involved combining two paper maps into one new paper map, using tracing paper and light tables. This technique is time consuming and inaccurate when undertaken by the untrained hand. The expert may still find the process time consuming and considerably expensive. For these reasons, GIS is in its element in conducting overlay processes.

The simplest GIS overlay combines data—spatial and attribute—from two layers into a new layer. In most vector overlay processes provided with GIS software, one of the layers is a polygon layer.

Combining two polygon layers is conceptually simple (Figure 9.8). The result will be a new polygon layer. The new layer will contain all spatial data from both input layers, and may therefore contain many new polygons. Attributes from both layers will also be retained.

Boolean operators are often used in vector polygon overlays. Examples of operators include AND, NOT and OR. Boolean operators are used to test whether a certain state or condition is true or false. An example of the use of Boolean operators in a query constructed using the data in Figure 9.8 may be: 'Overlay suburb and land-use layers to create a new layer showing suburb = east valley AND rural residential'. Only polygons that satisfy both of these criteria will be identifiable in the new layer.

Combining a line and a polygon layer is called a line-in-polygon analysis. This analysis literally determines which polygon each section of line overlays (Figure 9.9). For example, an overlay of a polygon layer of soil types and road locations will result in a line layer showing road locations with attached attributes from the soil layer. Lines are segmented wherever they cross a polygon boundary.

Combining a point and a polygon layer is referred to as a point-in-polygon analysis. This analysis determines which polygon each point overlays (Figure 9.10). Example layers in a point-in-polygon analysis may be points showing social security centres, and polygons depicting levels of unemployment. The user may wish to retain the data as points and to attach the unemployment data to the point location in a new layer.

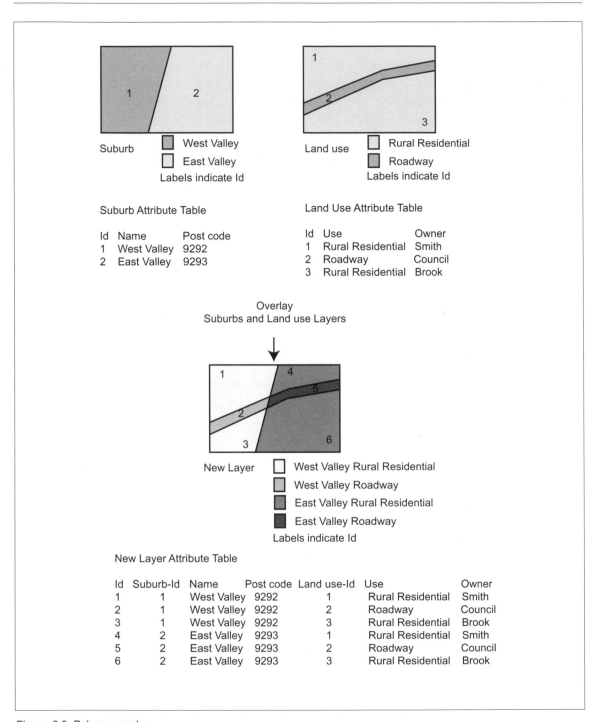

Figure 9.8 Polygon overlay

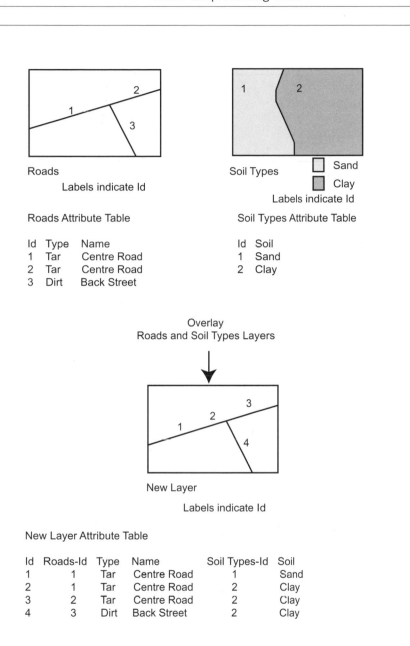

Roads
Labels indicate Id

Soil Types
Sand
Clay
Labels indicate Id

Roads Attribute Table

Id	Type	Name
1	Tar	Centre Road
2	Tar	Centre Road
3	Dirt	Back Street

Soil Types Attribute Table

Id	Soil
1	Sand
2	Clay

Overlay
Roads and Soil Types Layers

New Layer
Labels indicate Id

New Layer Attribute Table

Id	Roads-Id	Type	Name	Soil Types-Id	Soil
1	1	Tar	Centre Road	1	Sand
2	1	Tar	Centre Road	2	Clay
3	2	Tar	Centre Road	2	Clay
4	3	Dirt	Back Street	2	Clay

Figure 9.9 Line-in-polygon overlay

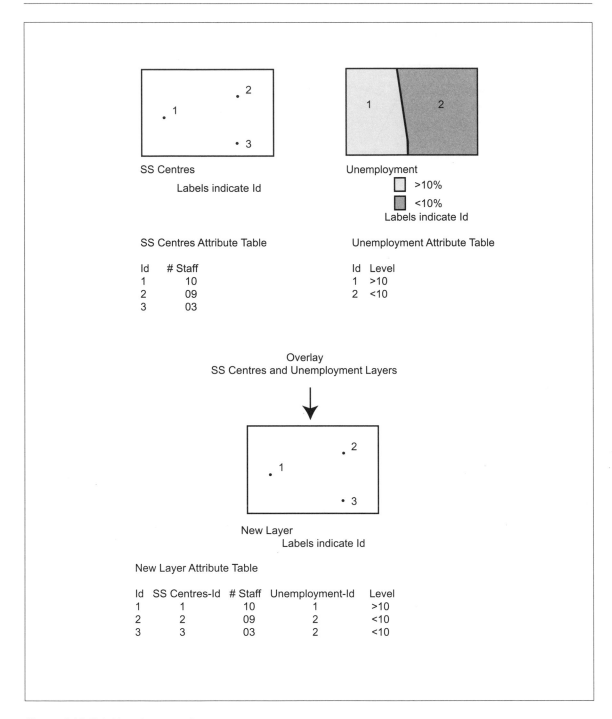

Figure 9.10 Point-in-polygon overlay

The spatial extent of the resultant layer can be limited to either of the input layers or a union of both (Figure 9.11). Some packages even allow the user to interactively define an area for overlay.

Figure 9.11 Overlay result spatial extent

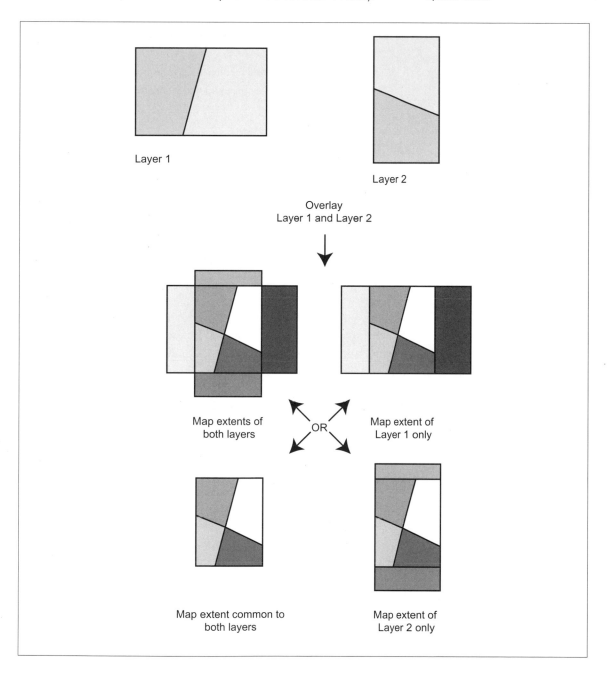

Layer 1

Layer 2

Overlay
Layer 1 and Layer 2

Map extents of both layers

OR

Map extent of Layer 1 only

Map extent common to both layers

Map extent of Layer 2 only

Vector geoprocessing tools: Comments from the workplace

The survey responses to this topic were varied. Some GIS users found these tools invaluable and a part of their everyday work.

> I use these commands and features a lot in my GIS work. Buffering is particularly important . . . I use clip and merge to produce larger maps.
>
> *Nicholas Welch, Planning Officer*

GIS users who are consultants, called into many departments for short-term work, found the use of vector geoprocessing tools rare in their line of duty.

> Most jobs I've done or applied for have few instances to use 'real' GIS tools like these—it seems that basic data entry and validation are the most crucial tasks in the industry at the moment.
>
> *Piers Higgs, GIS Consultant*

The most commonly used vector geoprocessing tools are those which have been tailored to a specific use and result in real savings in time and data purchase. Examples such as updating dynamic data sets are common.

Exercises: Geoprocessing Jarrahlea

Figures 9.12 and 9.13 contain examples of vector geoprocessing tools. Use these figures to answer the following questions.

Q1. Which two geoprocessing tools might have been used to arrive at Figures 9.12a and 9.12b?

Q2. Figure 9.12c shows the area within a buffer. This area is kept as clear as possible of flammable material such as bark and leaves. Which built structures have been buffered? What comment could you make regarding the spatial extent of the buffer of the property boundary?

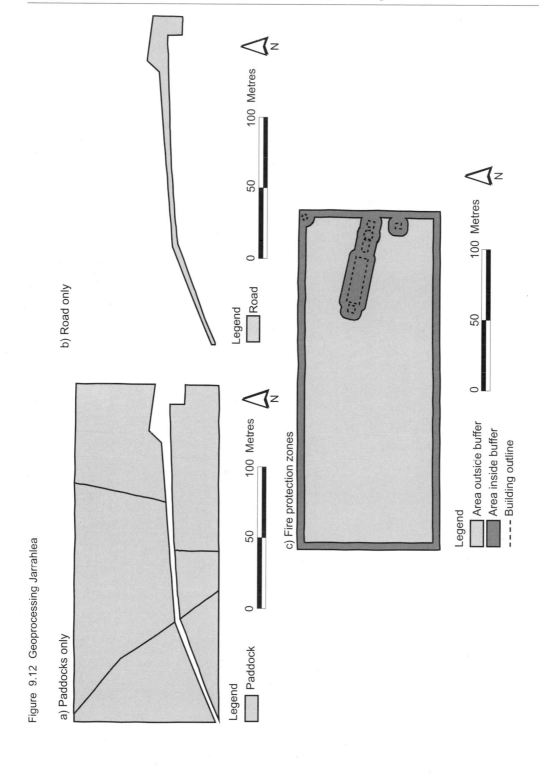

Figure 9.12 Geoprocessing Jarrahlea

a) Paddocks only

Legend
Paddock

b) Road only

Legend
Road

c) Fire protection zones

Legend
Area outside buffer
Area inside buffer
- - - - Building outline

Q3. Figure 9.13a shows an example of merging data. The data displayed show the amount of grass cover in each polygon in an early spring month. Use previous figures (such as Figure 1.5) to note the names of paddocks that have been merged.

Q4. Figure 9.13b is an update of Figure 7.5 after clearing land for a firebreak. Explain how the update has worked.

Figure 9.14 provides an example of a vector overlay. The overlay has combined attribute data from two layers, paddocks and marri trees, with the spatial data related to the marri tree locations.

Q5. Is it possible to determine that an overlay process has occurred simply by looking at the spatial data in Figure 9.14c?

Figure 9.13 More geoprocessing of Jarrahlea

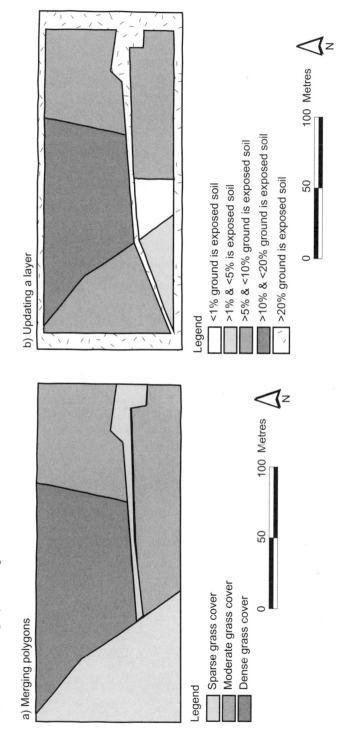

a) Merging polygons

Legend

☐ Sparse grass cover
▨ Moderate grass cover
▦ Dense grass cover

b) Updating a layer

Legend

☐ <1% ground is exposed soil
☐ >1% & <5% is exposed soil
▨ >5% & <10% ground is exposed soil
▦ >10% & <20% ground is exposed soil
▨ >20% ground is exposed soil

Figure 9.14 Layer upon layer of Jarrahlea data (vector)

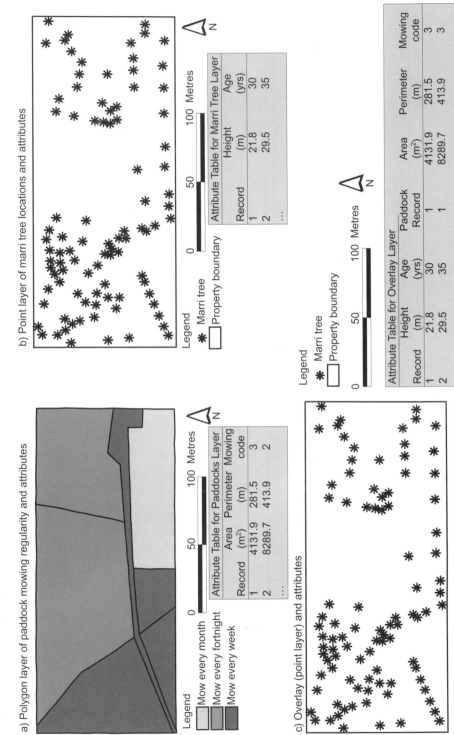

a) Polygon layer of paddock mowing regularity and attributes

Legend
- Mow every month
- Mow every fortnight
- Mow every week

Attribute Table for Paddocks Layer			
Record	Area (m²)	Perimeter (m)	Mowing code
1	4131.9	281.5	3
2	8289.7	413.9	2
...			

b) Point layer of marri tree locations and attributes

Legend
- ✳ Marri tree
- ▢ Property boundary

Attribute Table for Marri Tree Layer		
Record	Height (m)	Age (yrs)
1	21.8	30
2	29.5	35
...		

c) Overlay (point layer) and attributes

Legend
- ✳ Marri tree
- ▢ Property boundary

Attribute Table for Overlay Layer						
Record	Height (m)	Age (yrs)	Paddock Record	Area (m²)	Perimeter (m)	Mowing code
1	21.8	30	1	4131.9	281.5	3
2	29.5	35	1	8289.7	413.9	3
...						

CHAPTER TEN:

Raster Geoprocessing Tools

Raster geoprocessing can be undertaken as point (cell by cell), **neighbourhood**, or zonal analyses. These operations can be built into complex tools for the purpose of analysing and manipulating continuous data. Operations used to create data, such as drainage, slope and aspect, discussed in the following chapters (particularly Chapter Twelve), draw on many of these raster geoprocessing tools.

Raster geoprocessing

Geographical data represented by a raster structure are limited by the resolution of the cell and the need to derive topology. Raster data does, however, allow the use of a variety of geoprocessing tools.

Raster geoprocessing alters pre-existing data to create derived, new data, much like vector geoprocessing. Many of the operations that may be classified as raster geoprocessing are addressed in Chapter Twelve, as they relate more specifically to landscape analysis tools. This chapter will introduce point, neighbourhood, and zonal operations for raster data. Armed with this toolbox, the GIS user can achieve a range of operations similiar to those discussed for vector data in Chapter Nine.

Point operations in raster geoprocessing

Point operations refer to operations that calculate a new value for a raster grid cell on a cell-by-cell basis (Figure 10.1). The most conceptually simple raster geoprocessing involves grid layers that directly overlay each other. This means that all the rows and columns in grid layers must line up, cells must be of the same size, and the grid layers must occupy the same spatial extent (Figure 10.2).

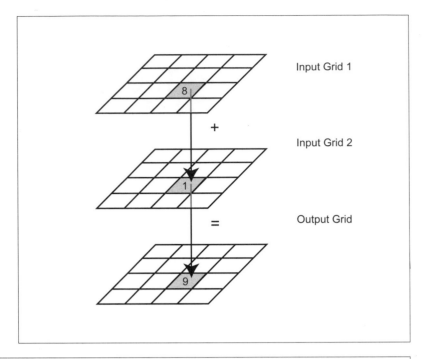

Figure 10.1 Raster point geoprocessing

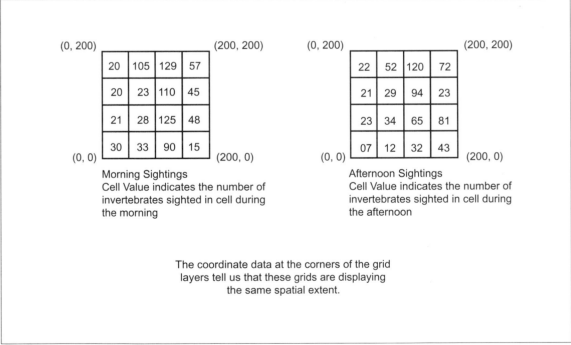

Morning Sightings
Cell Value indicates the number of invertebrates sighted in cell during the morning

Afternoon Sightings
Cell Value indicates the number of invertebrates sighted in cell during the afternoon

The coordinate data at the corners of the grid layers tell us that these grids are displaying the same spatial extent.

Figure 10.2 Raster data spatial extents

There are, of course, ways in which raster geoprocessing may be undertaken without these requirements. For example, a grid layer may be generalised to a larger cell size or resampled to a smaller cell size. The user may even define a specific area of each grid to be overlaid, such as the common map extent, rather than entire grids.

The principles supporting raster overlay, a point operation in raster GIS, are often termed **grid algebra** (also called map algebra), or Boolean algebra.

A mathematical statement can be used to describe the manner in which grid layers are to be combined. For example:

$$grid1 + grid2 = grid3$$

$$invertebrates\ sighted\ (AM) + invertebrates\ sighted\ (PM)$$
$$= total\ invertebrates\ sighted$$

The mathematical symbol indicates how the GIS software will merge the inputs (cell values) to create an output grid layer (Figure 10.3). These are commonly, but not exclusively, addition, subtraction, multiplication, or division. Alternatively, the GIS user may wish to consider input cell values (one from each input grid layer) and select one of the original values for the new grid layer cell. The selection may be based on a need to find the maximum or minimum value. Another possibility is that the GIS users could require an output that averages the input numbers. Examples of these forms of grid algebra are shown in Figure 10.4.

The selection of the best method for combining data from multiple grids into one new grid requires knowledge of the original grid cell values. In most cases a simple form of grid algebra is not appropriate (Figure 10.5). An alternative way to undertake raster overlay is to use a code system to fully define unique new cell values for each possible combination of input grid cell values. This helps to avoid the potential problems identified in Figure 10.5. For example: if the cell in grid one has a value of 2 (wildlife reserve) and the same cell in grid two has a value of 1 (indicating one sighting of a rare species), put a value of 8 in the resulting grid, conservation values. The conservation layer data dictionary should then note that a value of 8 is a cell with one sighting of a rare species in a wildlife reserve.

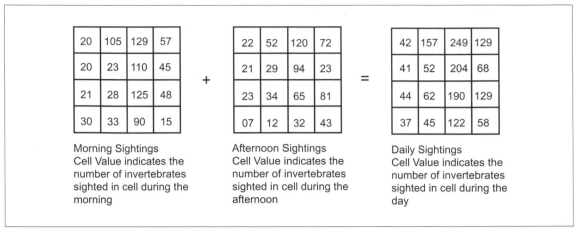

Figure 10.3 Grid algebra example

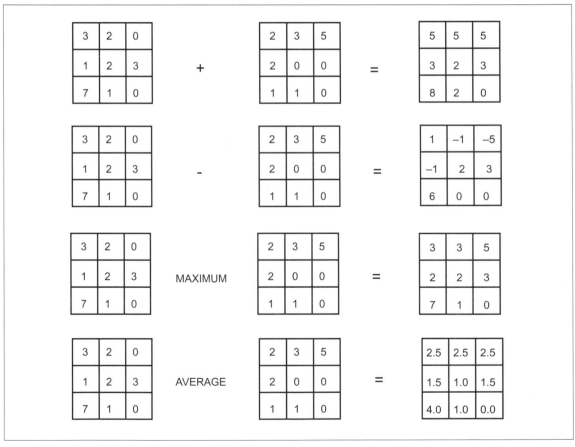

Figure 10.4 Grid algebra operators

Land Cover
0 = lake
1 = cleared
2 = wildlife reserve

1	2	0
1	2	2
1	1	2

Rare Species Sightings
0 = no sighting
1 = one sighting
2 = two sightings

2	1	0
1	2	0
1	1	2

Combine data in
Land Cover and Rare Species Sightings
to determine conservation values

Solution 1
Using +

3	3	0
2	4	2
2	2	4

Land Cover/Rare
Species Sightings
0 = lake/no sighting
2 = lake/two sightings or
    cleared/one sighting or
    wildlife res./no sighting
3 = wildlife res./one sighting or
    cleared/two sightings
4 = wildlife res./two sightings

Solution 2
Using a coding scheme

6	8	1
5	9	7
5	5	9

Land Cover	Rare Spp. Sightings	Result
0	0	1 = lake/no sighting
0	1	2 = lake/one sighting
0	2	3 = lake/two sightings
1	0	4 = cleared/no sighting
1	1	5 = cleared/one sighting
1	2	6 = cleared/two sightings
2	0	7 = wildlife res./no sighting
2	1	8 = wildlife res./one sighting
2	2	9 = wildlife res./two sightings

Figure 10.5 Grid algebra outcomes

Boolean algebra uses operators, such as AND, NOT, and OR to test whether a certain state or condition is true or false. For example, we may want to identify raster cells that have been experiencing heavy traffic flows at any time of the day (morning or afternoon). A Boolean-based query may read:

morning traffic = high OR afternoon traffic = high

The query searches for cells that have a high code in either the morning traffic grid layer or the afternoon grid. Usually, putting a value of one (1) in the corresponding cell of the new grid layer (daily traffic) indicates a true response. A false response, i.e. cells that do not satisfy the criteria, would receive a zero (0) value in the new layer. Strictly speaking, this is raster geoprocessing: altering pre-existing data to create new layers, as well as a form of overlay, combining data, select data in this case, into a new layer.

Neighbourhood operations in raster geoprocessing

Neighbourhood operations involve the creation of new data for a cell based on a knowledge of the surroundings, i.e. on neighbourhood values. The user can define a neighbourhood shape and size. Figure 10.6 shows common types of neighbourhoods, e.g. a three-cell by three-cell square.

Some example uses for neighbourhood raster geoprocessing include finding the largest or smallest surrounding cell value, determining the average value, summing all values, or identifying the number of unique values (variety) within a neighbourhood (Figure 10.7). Basic neighbourhood geoprocessing can be used to achieve higher order analyses, such as filtering or smoothing data, building landscape models, and determining flow across a surface (as discussed in Chapter Twelve).

Zonal operations in raster geoprocessing

One raster grid can be used to define zones for the geoprocessing of another grid layer. The first grid, used to define the shape, size and

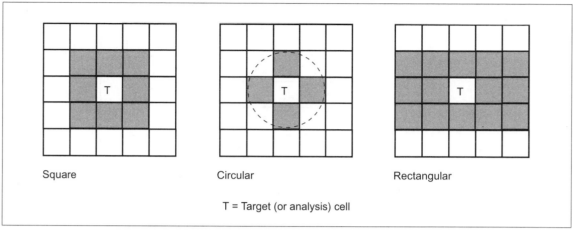

Square Circular Rectangular

T = Target (or analysis) cell

Figure 10.6 Grid
neighbourhoods

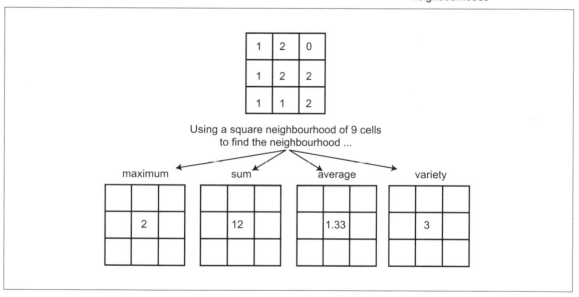

Using a square neighbourhood of 9 cells
to find the neighbourhood ...

maximum sum average variety

2 12 1.33 3

location of zones, will have cell values indicating unique zonal areas (Figure 10.8). These zones are not necessarily contiguous. The second grid contains the data of interest. The operations available in a zonal context are similar to those listed above for neighbourhoods, e.g. finding the largest or smallest grid cell value in each zone, finding the average or sum of all cell values in a zone, and determining the variety of values in each zone (Figure 10.9). The result (value) of a zonal operation is placed in each cell of that zone.

Figure 10.7 Example
neighbourhood analyses

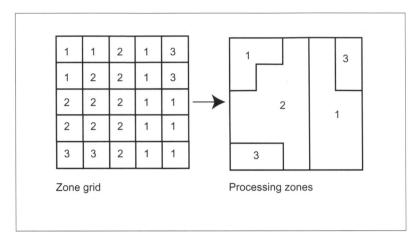

Figure 10.8 Grid zones

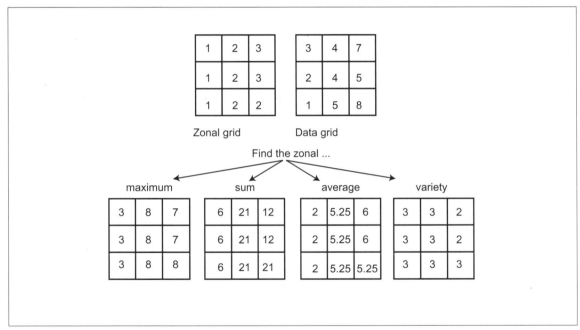

Figure 10.9 Example zonal analyses

Raster geoprocessing: Comments from the workplace

The survey respondents indicated that geoprocessing and overlay, raster and vector, are powerful tools that can independently justify the purchase of a GIS for some Australian organisations. Vector analysis and overlay is, however, more common than raster analysis and overlay. One respondent commented that this was due to the traditional forms of analysis and overlay being vector based and suggested that as GIS users realise the advantages of illustrating truly continuous data as raster grid layers, grid algebra will grow in acceptance. Another respondent believed that cost was the main prohibitive factor:

> When they tell me how much they're prepared to pay on software and hardware then it usually comes down to just vector [overlay].
>
> *Nick Middleton, GIS Consultant*

Those survey respondents whose data necessitates the use of grid algebra find it a flexible and adaptable tool. The quote below represents a good summary of the opinions of raster overlay supporters.

I have used more raster overlay than vector overlay. This is due to the fact that most of the modelling I have conducted has been using continuous data. I have used raster overlay for a large variety of tasks including predicting flooded areas due to the damming of a river; implementing the Universal Soil Loss Equation within a GIS; modelling surface hydrology; image interpretation (using band ratios of Landsat TM images); and determining the volume of flood water for given flood depths (in order to assist in the planning of an appropriate drainage system).

> I try to avoid vector overlay, especially with complex datasets, due to the large processing time. It is usually possible to get around having to use vector overlay.
>
> I also prefer raster overlay to vector overlay because most GIS that I have worked with allow you to use some form of map algebra. Therefore you are not limited to the [simplistic] vector overlay tools provided by the GIS software.
>
> *Narah Stuart, GIS Modeller*

Exercises: Layer upon layer of Jarrahlea data

Figure 10.10 is an example of a raster overlay. In this case the figure shows an addition of two data layers: one layer contains the soil moisture levels in summer, the other shows the soil moisture levels in winter. The combination of the two layers allows an investigation of the annual variability of soil moisture.

Q1. The cell values in the two grid cell layers (Figures 10.10a and 10.10b) have different ranges (10–20, and 1–3). How has this helped in the overlay process?

Q2. Some of the cells in Figure 10.10c are located over a water irrigation line that is only used in summer. What value has been given to these cells?

Figure 10.10 Layer upon layer of Jarrahlea data (Raster)

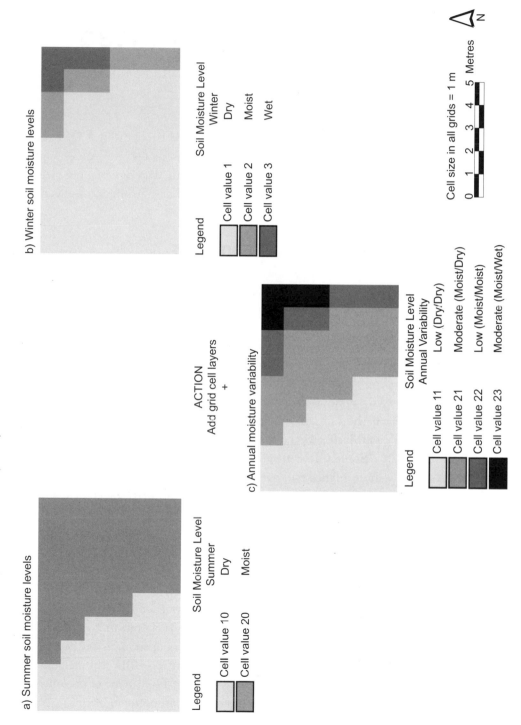

a) Summer soil moisture levels

Legend   Soil Moisture Level
          Summer

Cell value 10   Dry

Cell value 20   Moist

b) Winter soil moisture levels

Legend   Soil Moisture Level
          Winter

Cell value 1   Dry

Cell value 2   Moist

Cell value 3   Wet

ACTION
Add grid cell layers
+

c) Annual moisture variability

Legend   Soil Moisture Level
          Annual Variability

Cell value 11   Low (Dry/Dry)

Cell value 21   Moderate (Moist/Dry)

Cell value 22   Low (Moist/Moist)

Cell value 23   Moderate (Moist/Wet)

N

Cell size in all grids = 1 m

0   1   2   3   4   5   Metres

CHAPTER ELEVEN:
Analysis—Advanced

Network and proximity analyses are two powerful GIS tools that appear to be under-utilised. The reasons for the lack of application may lie with availability of appropriate data or ease of use within the software. These are tools, however, that may offer cost-effective solutions to certain problems.

More advanced analysis

The major portion of this book deals with the basics of GIS. The next few chapters, however, consider more complex analyses available within the GIS environment. These include network and proximity tools (Chapter Eleven), landscape analysis tools (Chapter Twelve), and modelling (Chapter Thirteen). The discussion regarding these tools will supply the reader with knowledge of aspects of GIS analysis and higher level modelling. This text presents only one method for arriving at specific results. Be aware that there are many alternative ways of performing the tasks discussed.

Some GIS software packages are designed specifically for one of these higher purposes rather than the tasks mentioned in previous chapters. These packages tend to be tailored to specific clients' needs and specific types of data.

Proximity analysis

Proximity analysis generally involves some kind of spatial search through a user-defined area (neighbourhood), and a measurement or calculation. It may involve a buffer operation to create a vector neighbourhood, or use a raster neighbourhood (as defined in Chapter Ten) and an addition to sum all entities within the neighbourhood. An example proximity query may be: 'How many optical dispensaries are located within one kilometre of an optician?'

Using vector data as an example, this would involve buffering optician locations to one kilometre to create a new layer. The new layer defines the neighbourhood (Figure 11.1). Next, this new layer would be used in an overlay with the optical dispensaries layer (point-in-polygon overlay). Finally, the GIS would need to add up the number of points in the overlay result that are inside the neighbourhood.

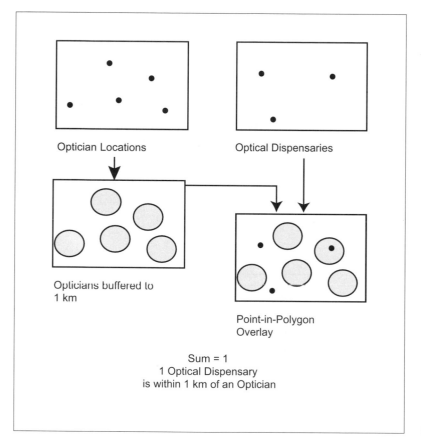

Figure 11.1 Proximity analysis

Another example may be the need to determine whether crime sites are clustering in space. Proximity analysis would involve mapping the spatial distribution of all crime sites, and then, for each site, determining the distance to the nearest neighbouring crime site. A statistical cluster analysis would then query whether the sites were in fact occurring closer to each other than expected in a random distribution.

The most commonly asked questions of proximity analysis are:

How far/close is a feature of type 'A' from/to another feature of type 'B'?

How many features of type 'A' occur within a given distance of a feature of type 'B'?

What is the closest/furthest feature of type 'A' from another feature of type 'B'?

Networks

Network GIS is based on connected and continuous line work. The lines may be roads, rail lines, shipping routes, bird migration paths, or any route that can be said to guide or transmit the flow of an entity (Figure 11.2).

Qualities such as cost, speed, time, impedance, direction, road surface condition, and traffic volume can be attributed to any unique section of a line.

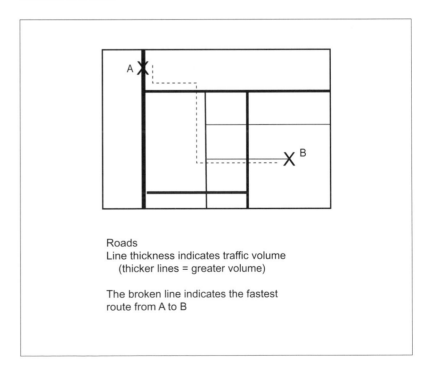

Roads
Line thickness indicates traffic volume
(thicker lines = greater volume)

The broken line indicates the fastest
route from A to B

Figure 11.2 Network analysis

Modelling the flow of an entity through a network along these linear features enables calculations such as the cost, length, or time taken for a trip from node A to node B along line sections c, d, e, and f.

Comparisons of different paths throughout a network can help to determine the quickest, cheapest, safest, shortest, or best path from node A to node B. This, in turn, allows analysis of accessibility, or the sum measure of how 'reachable' a number of alternative locations are from one identified location.

Some GIS packages are specifically designed for this type of analysis; others include network analysis as a separate module.

Coupling the network tool with the proximity tool has created a valuable GIS toolbox. In the transport and utility fields this has led to significant increases in efficiency, safety and profitability.

Advanced analysis: Comments from the workplace

Many GIS users in the emergency services, utility providers, and transportation industry have embraced the use of these two tools. It is fair to say that, globally, fields remain in which these capabilities are not being used to their full extent. The main reason for the lack of uptake of these tools, as indicated in survey responses, is the time required for the pre-processing of data in order to make the tools effective and efficient.

The amount of time needed to prepare the data and run the processes in GIS must be justifiable in terms of the complexity of the data. Also, in many cases respondents did not find proximity analysis tools intuitive or easy to use.

> In terms of proximity analysis, when it's as easy for me to get the GIS to see what's close to something as it is for me to do it visually, then I'll use the GIS.
>
> *Piers Higgs, GIS Consultant*

Coding the line work with attributes (such as direction and impedance) that are required for network analysis is time consuming, especially in a large network.

I've only used networks in academia. The data we usually deal with outside academia are not of a standard that can be used with network analysis—too much time has to be spent cleaning the data to make it worthwhile.

Piers Higgs, GIS Consultant

These tools do become indispensable in situations where the investment in creating data for network and proximity analysis is justifiable.

Networks are useful when trying to look at serviceable areas. This is [crucial] to my job in the fire department. Network analysis is used to determine which stations can service certain areas inside certain time limits.

Nicholas Welch, Planning Officer

Research will continue to identify application areas where adaptations of these tools are feasible and appropriate.

Exercises: Proximity analysis amongst the tea-trees and networking with Jarrahlea

Figure 11.3 displays data that could be used in both a proximity analysis and in a network study.

Q1. Which features could be relevant in a network analysis? Give an example of an appropriate research question for each network feature you identify.

Q2. Exploratory data analysis suggests that a proximity analysis could be employed to test for clustering patterns in one of the layers. Examine the figure and suggest the most obvious proximity analysis for the data.

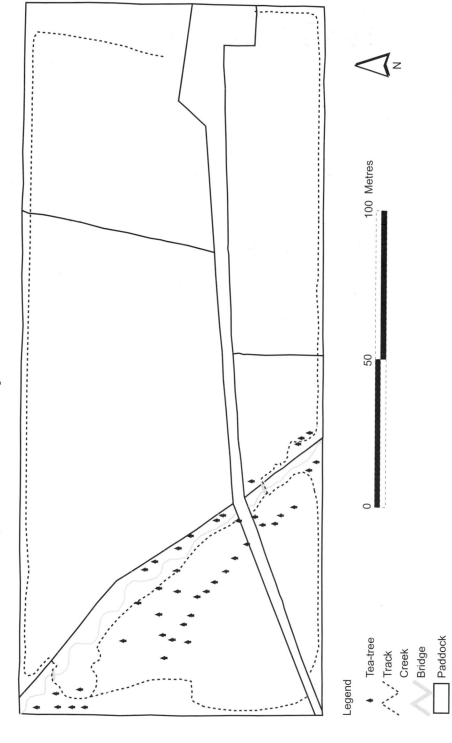

Figure 11.3 Proximity analysis amongst the tea-trees and networking with Jarrahlea

Legend

Tea-tree
Track
Creek
Bridge
Paddock

0   50   100 Metres

N

CHAPTER TWELVE:
Landscape Analysis

The use of **digital elevation models** in GIS has been popular. The third dimension, usually elevation, is stored as an attribute rather than as spatial data. Digital terrain models such as slope, aspect, convexity, concavity, viewsheds, profiles, volume estimation and drainage models are derived from digital elevation models.

Digital elevation models

A digital elevation model **(DEM)** is the GIS representation of the rises and falls in elevation across a landform in the real world, which varies in three dimensions rather than two (Figure 12.1). DEMs may be used as a basis for terrain analysis, or simply as a visualisation tool.

With the advent of virtual reality and similar computer visualisation techniques, there is a growing demand for viewing the GIS world

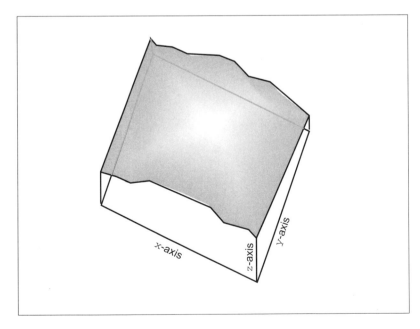

Figure 12.1 The third dimension

in three dimensions (3D). Users want to see the world displayed on the screen as it would look in real life. Elevation data, usually stored as contour lines, are the most widely used GIS building blocks in the creation of two and a half dimensional (2.5D) GIS landscapes.

Why do GIS users refer to these displays as 2.5D? Although the view may appear to be 3D on the screen, it is termed 2.5D as the height, or elevation, is stored as an attribute rather than as spatial data. This means that overhangs and features that really do have more than one elevation value for one *x*–*y* location are problematic (Figure 12.2).

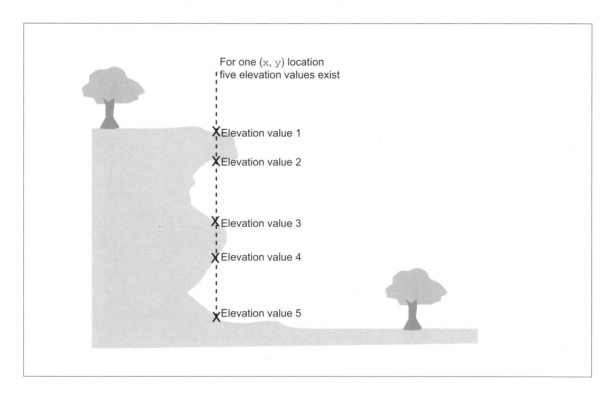

Forms of data other than elevation contours can be used to create these 2.5D views. Spot heights or surveyed profiles can also be used. Alternatively, a 2.5D display may be created using data that is not related to elevation. Any spatial data with an attribute that can be displayed as rises and falls may form the 0.5D. Consider a 2.5D 'landscape' of population density, traffic volume, temperature, or precipitation.

Figure 12.2 Multiple elevation values for one spatial location

The progression from contour (line) or point data into a form of 2.5D data can be undertaken in a variety of ways. The most common ways use triangulation or **interpolation** and gridding of data. Both methods usually employ point data as the basic building element. Non-point data (contour lines or polygons) can be transformed into points in order to build a DEM.

Triangulated irregular network (TIN)

A **Triangulated Irregular Network (TIN)** is an elevation surface of triangular elements (Figure 12.3). A TIN is structured much like a polygon layer. It contains topology and each triangle could be considered as a three-sided polygon.

TINs can be effectively shaded using attribute values such as aspect and slope of each triangular facet (Figure 12.4).

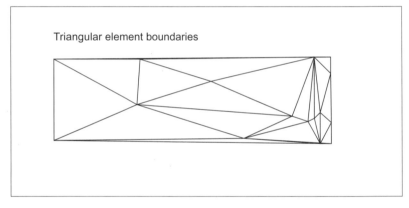

Triangular element boundaries

Figure 12.3 A TIN

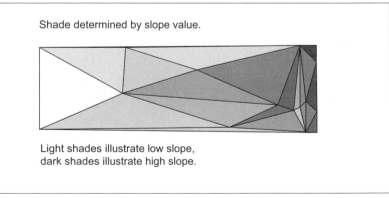

Shade determined by slope value.

Light shades illustrate low slope,
dark shades illustrate high slope.

Figure 12.4 A shaded TIN

The use of a TIN is advantageous in the display of 2.5D land-scapes with abrupt changes, such as cliffs and gullies, as the triangle bases can be aligned with these features. TINs are also ideal for landscapes that have both large areas of very low variability in elevation (flat areas) and areas of rapid change (steep mountains). Small triangles can be clustered around the data-rich, highly variable areas, and large triangles can be used to represent regions of very little undulation.

The creation of a TIN is computationally expensive, and the orig-inal point elevation data are often thinned prior to processing a TIN. The mass of point data can be thinned using a variety of methods. The most common methods identify how important each and every point is to correctly representing variation in elevation. The GIS user defines an acceptable generalisation level, or a threshold num-ber of points, to differentiate the data to be retained from those to be discarded. Data around areas of great changes in elevation, such as cliffs, become relatively important points, whereas data points on plains become relatively unimportant.

Once the important points have been identified they are linked into triangles. The most common way to form triangles is a tech-nique called Delauney triangulation. Delauney triangulation states that any three points form a triangle if the circumcircle of the formed triangle does not contain any other point within its interior (Figure 12.5).

Grid DEMs

A **grid DEM** is a grid cell layer in which each cell value is an eleva-tion (Figure 12.6). Grid DEMs can be cleverly shaded to provide extremely effective 2.5D displays. Many GIS users find the grid DEM preferable to a TIN, as it is simple to incorporate grid DEMs into any grid-based modelling and analysis procedure.

A grid DEM is usually created from a point layer, through inter-polation between the known values to create a continuous surface. A grid DEM constructed using interpolation is also visually preferable to a TIN in data-sparse areas, as the TIN triangular elements would be large and obvious.

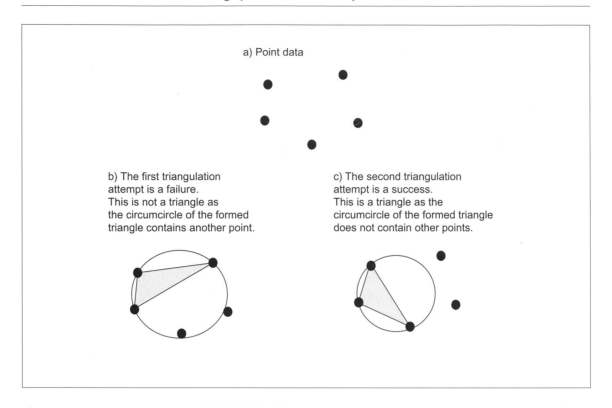

Figure 12.5
Delauney triangulation

20	21	22	23	24	35	39	47
20	21	22	25	27	29	37	45
22	22	22	25	23	34	48	53
22	23	24	26	29	38	39	43

Each cell contains an elevation value in metres
above sea level.

Figure 12.6 A grid DEM

Grid DEMs are easily processed and passed between packages
(particularly imagery software) and have been the most popular form
of DEM integrated with imagery.

There are situations where the use of the grid DEM rather than a
TIN is not advantageous. A grid DEM could have a large amount of

data redundancy in areas where terrain is low in variation. The grid cell size is constant across a surface, whereas the TIN element can alter in size between data-rich (small triangles) and data-poor (large triangles) locations (Figure 12.7).

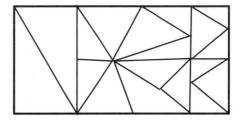

20	21	22	23	24	35	39	47
20	21	22	25	27	29	37	45
22	22	22	25	23	34	48	53
22	23	24	26	29	38	39	43

Each cell must contain a value.
There are 32 separate entities (cells) stored, each with an elevation value.

This TIN has 18 elements created from the point data set. The areas of low variability have large triangles and the areas of high variability have many small triangles. This means the TIN stores more important data in a lower number of entities (triangle/cells).

The choice of cell size in grid DEMs is critical in the representation of features. Small but important variations, such as a deeply incised river, could easily be lost through generalisation with an inappropriately large cell size.

Grid DEMs are ideal for visualisation in landscapes that do not have abrupt changes such as cliffs, and data that are sampled almost regularly across a landscape.

Figure 12.7 A DEM and a TIN: data redundancy

Lattices

A further DEM storage and display method involves the use of a **lattice** of data (Figure 12.8). A lattice is similar to a grid formation with the exception that data is held at nodes where column and row lines meet, rather than within the area of a cell. The empty areas between nodes and the connection of data-important nodes make this display resemble a fish net. Raising or lowering these fish net nodes to relative elevation levels gives a visually pleasing representation of the real world.

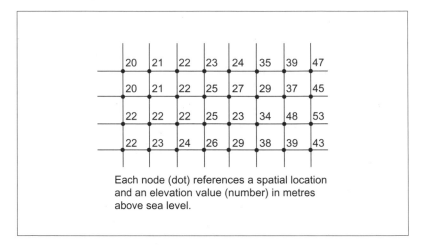

Each node (dot) references a spatial location and an elevation value (number) in metres above sea level.

Figure 12.8 A lattice DEM

Digital terrain models

The term **digital terrain model (DTM)** is often used synonymously with DEM. It is more correctly applied to any derivative of a DEM. These derivatives include slope, aspect, convexity or concavity maps, **viewshed**s, profiles, volume estimations, and drainage networks and basins.

There are many different ways for arriving at DTMs. Remember that these pages are supplying the reader with only one example of a multitude of techniques and tools available to the GIS user.

Slope and aspect

A measure of slope is a measure of the gradient of the land. The gradient is the maximum rate of change of elevation across a triangular feature or grid neighbourhood. The aspect is the direction, or bearing, of the maximum slope (Figure 12.9). Aspect is generally given as the compass direction of this maximum rate of change. Uses for slope and aspect studies may include studying flora and fauna habitat preferences, or finding flat land with a protected aspect for locating building sites.

Further derivatives of these measures are the convexity and concavity of the landscape. These can aid studies involving volume estimation, such as the volume of earth needed to level a sloping section

The centre cell is the cell of interest.

The slope (gradient) is calculated using the mathematical equation tan (slope) = change in elevation/run.

The change in elevation is the maximum difference between the centre cell and surrounding cells (3 m in the example).

The run will be the distance from the centre of the centre cell to the centre of the other cell. This can be calculated with a knowledge of cell size.

If our cell size is 1 m, in the example, run = ($\sqrt{2}$) or 1.414, therefore tan (slope) = 3/1.414, and slope is 64.761 degrees.

The aspect is defined by the direction of the maximum rise or drop from the centre cell (south-west in the example).

Figure 12.9 Slope aspect (a grid neighbourhood example)

of land (decreasing concavity). Concavity/convexity studies and volume estimation are particularly important in the mining and engineering fields.

Viewsheds and line of sight mapping

Viewsheds delimit all the areas of the ground which can be seen from a user-defined point or area (Figure 12.10). Another, simpler derivative is the line of sight map, which generally involves the user defining an origin (standing point) and a focus (point towards which the viewer is looking) (Figure 12.10). Conducting

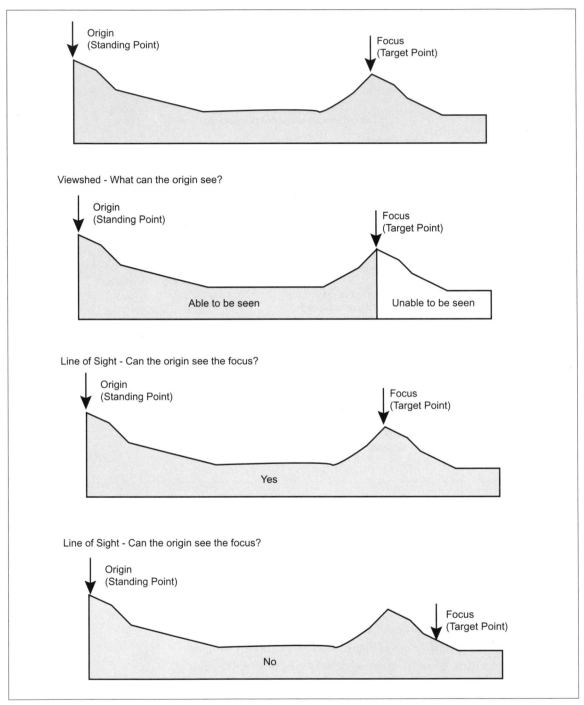

Figure 12.10 Viewsheds and line of sight mapping

either of these analyses without GIS would be time consuming, particularly in an undulating or complex landscape.

Uses of these tools are predominantly linked with 'what-if' visualisation studies, such as determining whether the tourists visiting a lookout will see the future felling site in a forest, or the impact of a new, proposed building on the view from an existing structure.

Drainage models

Once a DEM has been created it is a relatively simple task to create a number of layers of information related to drainage (Figure 12.11). A search for the landscape highs and lows can be used to label ridges and drainage lines. High points, when connected, will form a **drainage divide** and low areas will often represent watercourses,

Figure 12.11 Drainage models

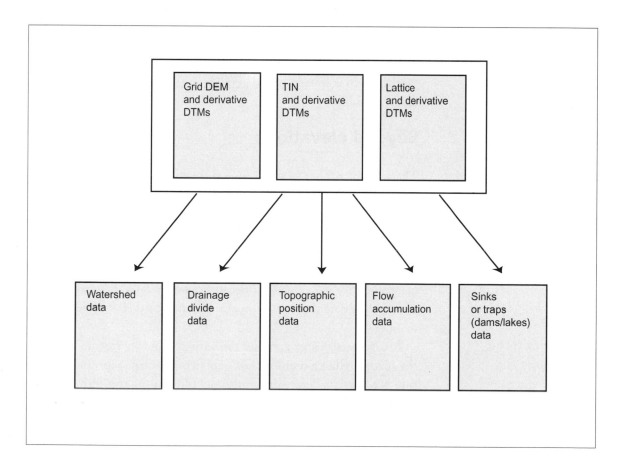

where overland flows converge or a river may flow. Most GIS models of **flow accumulation** use a grid DEM and the derived slope aspect to determine the path of a drop of water placed anywhere in a catchment. The path determined is based on the assumption that the drop of water will move down-slope in a manner that uses least energy, i.e. under the influence of gravity, moving down the steepest slope. The flow accumulation figure given to a grid cell is equal to the number of cells in the grid layer that drain through that particular cell.

Most software packages come with the algorithms that allow the derivation of these **drainage model**s from a DEM. It is important to understand the data, the landscape, and the algorithm used by the GIS software in order to correctly use these products. Realise, for example, that these GIS algorithms do not consider factors that may be of great importance (such as geology, soil type, or vegetation) and rely entirely on slope and aspect.

Example uses of drainage models include modelling water flows across a surface after a storm event, approximating expected pollutant flows in a system, and determining where pooling occurs on a landscape surface.

Beyond elevation

There are a number of interesting diversions from the use of elevation data in the creation of 2.5D surfaces. Climatic data (such as precipitation and temperature) and census data (such as population density) have also been used. For example, a 2.5D surface can be used to show where population densities are relatively high and low. The 'slope' derivative may be used to define areas where there is a great disparity in population density over a relatively small spatial extent. Any data that is continuous in nature could be viewed in this manner.

A word of warning: consider the variation in the size and shape of the spatial collection unit (census collection district) for the population density data before drawing conclusions using this form of visualisation tool!

Landscape analysis: Comments from the workplace

Using the 2.5D modelling capabilities in GIS has been popular, although it appears that **imagery packages** have the edge in performance with features such as **fly-through**s, which allow the user to move through a landscape in real time. The success of the use of 2.5D modelling is largely dependent on the quality of the available software, hardware, and data. A slow machine will frustrate the user, software not supporting various elements of landscape analysis may result in generalised solutions, and the resolution of the data may be too poor to create a semblance of reality.

The majority of survey respondents had a use for landscape analysis. Some simply used the 2.5D display as a visualisation tool. Most, however, actually used the tools for modelling or analysis.

> I have used DEMs very extensively! Mostly for surface hydrologic modelling, i.e. defining catchments and basins. Also for determining areas too steep to be cleared/mined/farmed—as part of environmental impact assessments. I have also used lattices to determine volumes of soil or rock that would need to be stockpiled during quarrying operations. I think that DEMs have a much wider application than most users are aware of and can be used to represent many kinds of continuous data, not only elevation. I found that non-GIS users are very impressed with data that is presented in 3D [2.5D]!
>
> *Narah Stuart, GIS Modeller*

Many of the survey respondents were aware of applications beyond elevation data.

> I have designed a [Graphic User] Interface using slope analysis for census data which produced reasonably promising results.
>
> *Nicholas Welch, Planning Officer*

Others emphasised the need to couple knowledge of concepts and practicality with experimentation.

A lot of 3D stuff gets done simply because it's possible. In many cases it doesn't have much scientific merit.

Werner Runge, GIS Project Officer and Consultant

Exercises: The highs and the lows of Jarrahlea

Figure 12.12 has three displays relating to the elevation data for Jarrahlea. Examine these three very different forms of data and answers the questions below.

Q1. Figure 12.12a is a grid DEM. The elevation data shows a general rise from the south-west corner to the north-east corner of the property. Can you trace where the creek runs through the landscape?

Q2. The TIN (Figure 12.12b) is shown as a collection of triangular elements. These elements could be shaded to give a diagram similar to Figure 12.12a. Try to explain why there are many small triangles clustered around the creek and larger triangles occurring towards the centre of the property.

Q3. The elevation is displayed as a lattice mesh in Figure 12.12c. The data relating to elevation are held at the nodes of the mesh. Explain how this view has been made to look three dimensional.

Figures 12.13a and 12.13b are example DTMs of Jarrahlea. Examine them closely and answer the following questions.

Q4. The aspect map, Figure 12.13a, displays the slope aspect of the property. Explain why aspect is a particularly difficult feature of a landscape to show using symbolisation or shading.

Q5. The slope map shows some interesting features. Explain the very steep slopes in terms of your knowledge of the property.

Figure 12.12 The highs and the lows of Jarrahlea (DEMs)

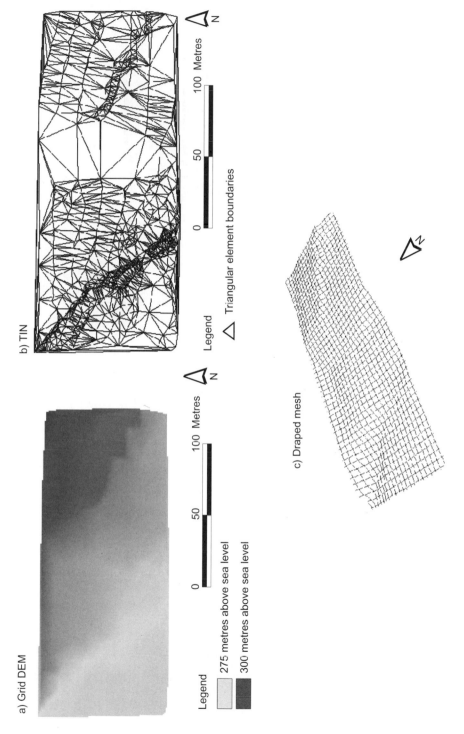

a) Grid DEM

Legend
275 metres above sea level
300 metres above sea level

b) TIN

Legend
△ Triangular element boundaries

c) Draped mesh

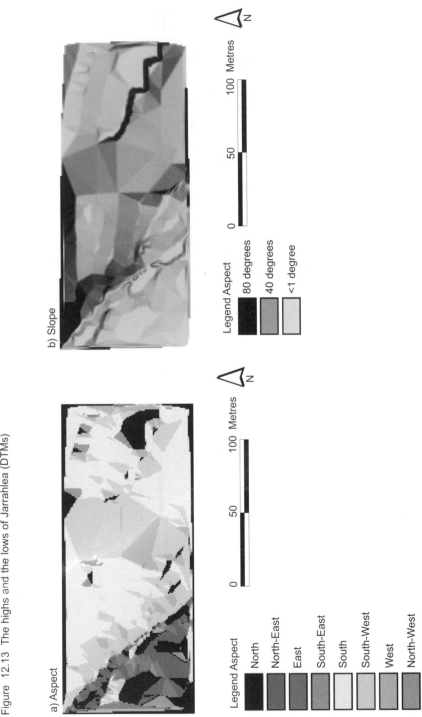

Figure 12.13 The highs and the lows of Jarrahlea (DTMs)

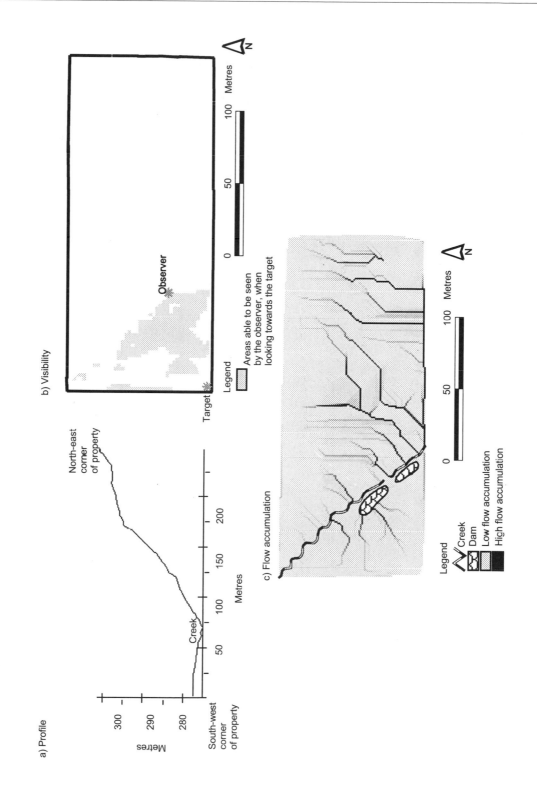

Figure 12.14 The highs and the lows of Jarrahlea (viewsheds and drainage)

Q6. Do you think that the aspect and slope maps were created from the grid DEM or the TIN? What evidence could you use to support this assumption?

Figures 12.14a and 12.14b contain results from a visibility analysis. The profile graph shows the changes in elevation a walker would experience walking in a straight line from the far south-west to the north-east corner of the property. The visibility map shows what an observer would see if they were positioned where the road crosses the creek, looking towards the gate at the south-west corner of the property. The final figure (Figure 12.14c) is a drainage model outcome showing the flow accumulation across the property.

Q7. Where (at what distance along the x-axis) is the steepest section of the profile?

Q8. Can the observer see the gate (focus) from where they are standing?

Q9. Where are errors occurring in the flow accumulation model? Can you suggest why they are occurring?

CHAPTER THIRTEEN:

Modelling

Modelling in a GIS environment is both flexible and efficient. GIS is able to handle many data types (both quantitative and qualitative) and sources (e.g. surveys and topographic maps), allowing complex modelling which was previously not attempted. These models may derive relationships, based on accepted formulae, built around available evidence, or tailored to specific unique conditions. Flow charts, and cartographic modelling in general, provide ideal tools for planning and describing GIS modelling. The discussion in this chapter centres on different classifications of models. Some of these classifications overlap in terms of the types of GIS operations used in the modelling and the desired results.

GIS and modelling capabilities

Modelling in GIS refers to a complex analysis involving many of the tools in the GIS toolbox. A GIS model usually has the aim of approximating a real world phenomenon or interaction of spatial and aspatial components in a computer-based (GIS) environment. A model may attempt to emulate a fully or partially understood process, be it a natural or cultural process. Many geographical and spatial modelling techniques have been adapted into GIS tools or operations. This chapter looks at the types of models common in GIS software and which are commonly employed by GIS users.

GIS models may be developed for a specific spatial extent or a specific phenomenon, and may be controlled largely by the user. Alternatively, a generic model may be adaptable and require minimal, or no, adjustment in order to be applied to a new data set.

The impetus for developing a model may be to try to explain the spatial phenomena occurring: 'How are the individual elements interacting to create a phenomenon?'; or 'Can we take what we have learnt about the current interactions and predict past or future scenarios?'

Assumptions form an important part of the GIS model. A model may have to assume that some factors are not going to change, are not exhaustible, or act as constants. Models can only approximate reality.

In GIS, modelling with natural or human-made phenomena can be explicative (What is . . . ?), predictive (What if . . . ?), or statistical (Compute the . . .). A complex situation may involve all three!

Explicative modelling

Explicative modelling uses the GIS toolbox to try to understand relationships between data, in order to describe what is apparent in the spatial and/or attribute data patterns (Figure 13.1). This is the most common use of GIS modelling. An example may be: 'Can I explain the distribution of this frog species in terms of this collection of variables (temperature, precipitation, soils, distribution of predator species, and permanent water locations)?'. Explicative modelling may form the first step in, or interact with, predictive and **statistical modelling**.

Predictive modelling

Predictive modelling involves a clearly understood situation or phenomenon, based on, or described by, a number of factors (Figure 13.2). One or many factors in the model are then altered in order to envisage how the greater system will respond, i.e. 'What if . . . ?'. Two hypothetical examples of predictive models are given below.

- *Physical Model:* soil erosion risk equation
  Variable: land use (change agricultural to residential land use)
  Question: How will the erosion risk change with the altered land use?
  Prediction (solution): initial increase in erosion risk by a factor of x during construction, followed by a decrease by a factor of y in risk due to soil compaction and application of pavements
- *Social Model:* spatial expansion of urban sprawl zone
  Variable: land subdivision approval rates
  Question: How would the act of lowering the subdivision approval rate affect the urban sprawl?
  Prediction (solution): initial decrease in urban sprawl by z%

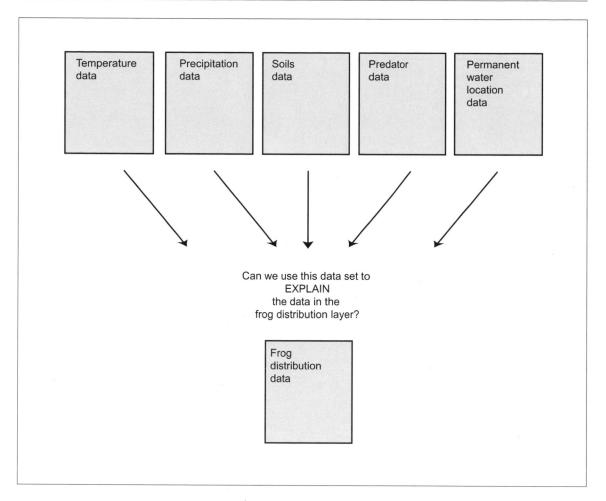

Figure 13.1 Explicative modelling

These models assume that there is an explanation for the current spatial trends based on a number of interrelationships between data. The question: 'What happens if factor x is altered?' is posed. The inputs are altered accordingly, the model run, and the GIS result is prepared for interpretation by an expert in the appropriate field.

Statistical modelling

Statistical modelling is usually based on quantitative data and an equation (Figure 13.3). For example, we may have a raster grid cell layer of land values and the proposed location of a new marina. Using a distance decay curve we may be able to mathematically

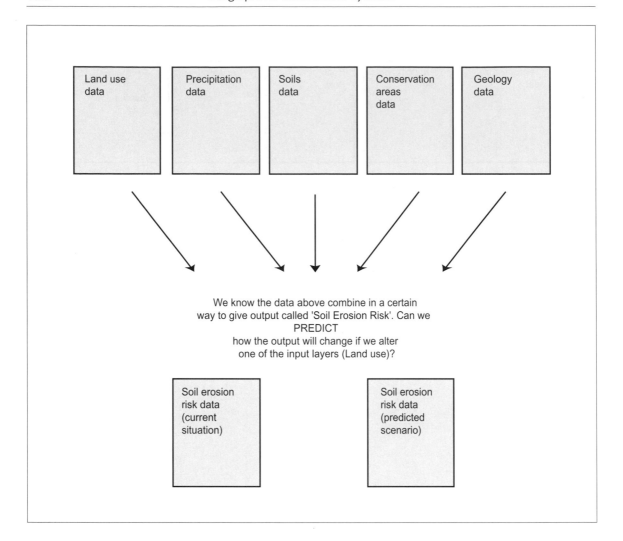

Figure 13.2 Predictive modelling

derive the added land value for surrounding properties, based on the attraction of locating houses close to the new marina.

Most GIS software packages offer simple functions to describe the data statistically. These simple statistical functions usually include the calculation of a mean, median, mode, variance, standard deviation, range, minimum and maximum. Many GIS software packages also provide tools enabling data to be viewed as histograms, bar charts, graphs, scatter plots and box plots. Some GIS software packages provide more advanced statistical tools, enabling principle component and factor analysis, cluster analyses, regression, and

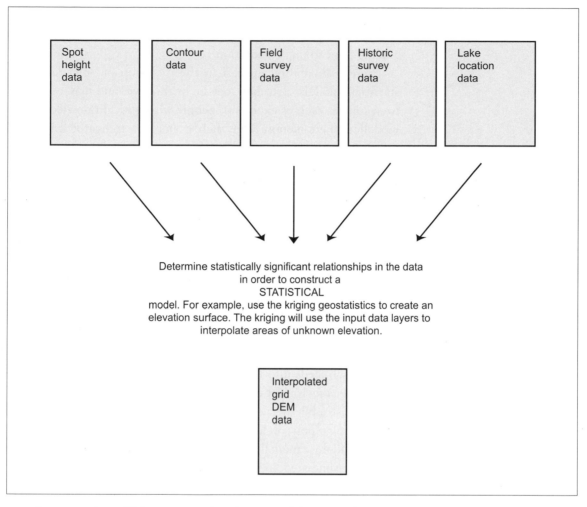

Figure 13.3 Statistical modelling

correlation analysis. If the statistical tools required for an analysis are not provided within the GIS software, users often link the GIS software with a separate statistical package, or otherwise employ programming skills in order to increase the functionality of the GIS software.

There are many other ways in which statistical and mathematical modelling concepts have been implemented in a GIS environment. Examples of these include the calculation of slope, as discussed in the previous chapter, interpolation, and surface fitting. **Kriging** is a prime example. Kriging is an optimal interpolation method that creates an interpolation based on the statistical description of variation of an attribute through space. Mathematically, it is quite a complex method for the novice and requires further background reading prior to implementation.

There is increasing interest in the inclusion of 'soft' data into statistical models. Soft data may be qualitative, and may be drawn from sources such as experts or people with a local knowledge of a modelled phenomenon. One such example is indicator kriging, a geostatistic.

Cartographic modelling

Reference has already been made in this chapter to the process of combining many of the tools and operations discussed in earlier chapters into a modelling structure. The combination of the techniques and operations, in an ordered manner, acting on data, to simulate a spatial decision-making process is the embodiment of **cartographic modelling**.

Cartographic modelling requires that each step, or operation, has a purpose. This purpose is usually to create a product, often a new layer, which feeds into the next or following steps. In the example given in Figure 13.4, the first operation is to find areas (census collection districts) that have a high proportion of young children. The result is a polygon layer that will feed into following operations.

Cartographic modelling also requires that a decision-making process be simulated. In the example, Figure 13.4, the question posed was: 'Is there a suitable location for a new child care centre?'. The result will be contained in the final model result (Result 5 in Figure 13.4).

The flow chart, shown in Figure 13.4, is a common way to represent cartographic modelling. Drawing a flow chart such as this, including the inputs, the operations, the linkages, and the output, is advantageous for a number of reasons. It constitutes a planning tool for the modelling exercise, illustrates how separate operations are linked together to achieve a result, and illustrates the method clearly to others.

Tomlin (1990) differentiated two types of cartographic modelling techniques: the descriptive and prescriptive modelling techniques. The *descriptive* model describes data patterns and may even attempt

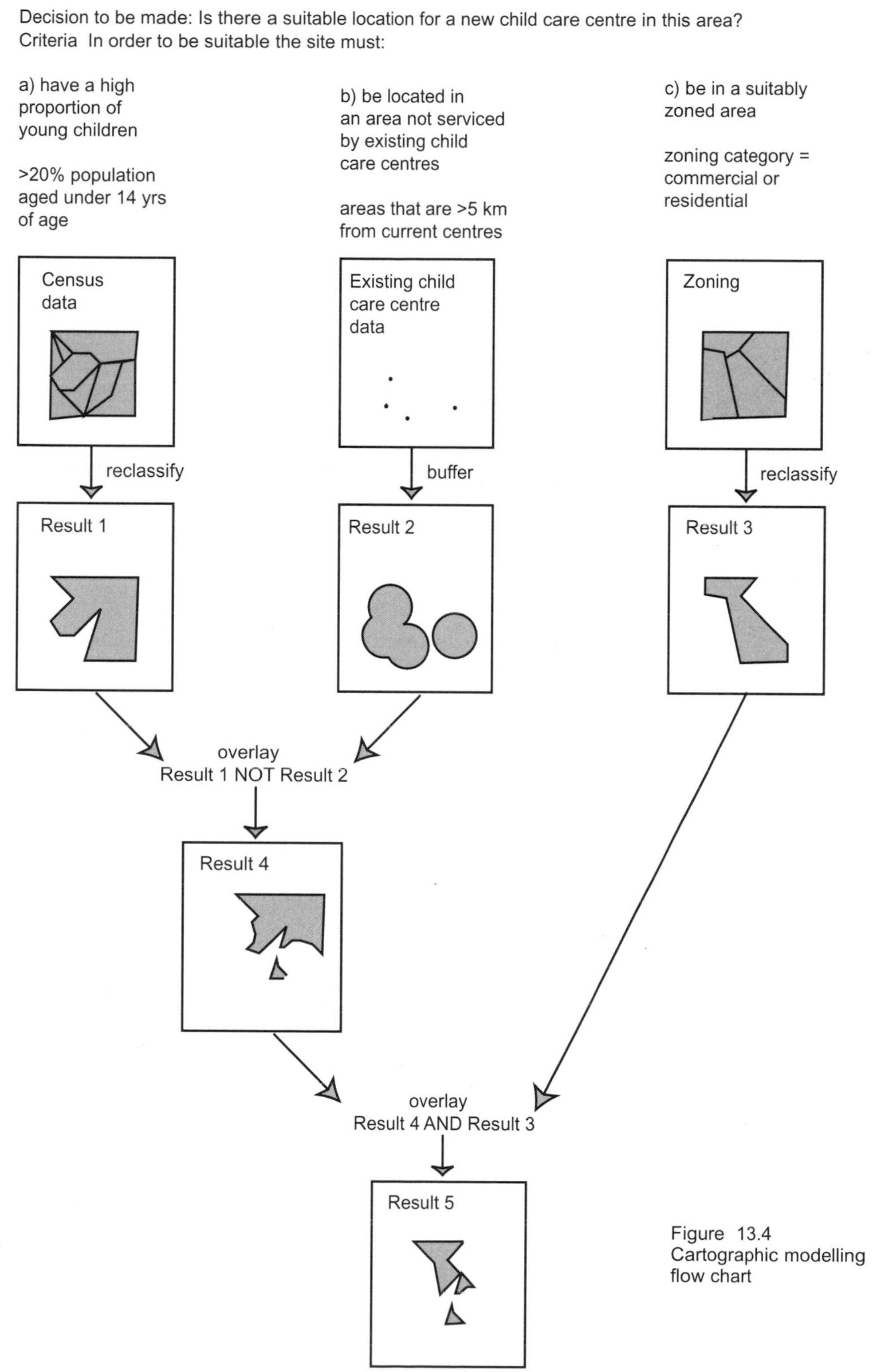

Decision to be made: Is there a suitable location for a new child care centre in this area?
Criteria In order to be suitable the site must:

a) have a high
proportion of
young children

>20% population
aged under 14 yrs
of age

b) be located in
an area not serviced
by existing child
care centres

areas that are >5 km
from current centres

c) be in a suitably
zoned area

zoning category =
commercial or
residential

Census
data

Existing child
care centre
data

Zoning

reclassify

buffer

reclassify

Result 1

Result 2

Result 3

overlay
Result 1 NOT Result 2

Result 4

overlay
Result 4 AND Result 3

Result 5

Figure 13.4
Cartographic modelling
flow chart

to explain associations. It may involve statistical models interacting with explicative models. The *prescriptive* model tends to be more complex than a descriptive model, and incorporates the statement of the problem, the generation of a result, and the evaluation of the result. They can involve explicative, statistical, and predictive elements. Figure 13.4 is an example of a prescriptive model. In prescribed modelling the problem statement is posed prior to commencement of the model implementation. The outline for the generation of a result (such as Result 5) is given diagrammatically. Once achieved, this result could be evaluated in terms of the criteria (items a, b, and c in Figure 13.4) and in terms of other aspects not included in the model. For example, the result (Result 5) may now be compared with available real estate.

GIS modelling: Comments from the workplace

There are many application areas in which Australians are at the forefront of international research in complex GIS modelling. When it comes to the level of the GIS user, however, most modelling being undertaken uses algorithms supplied by software companies.

The survey respondents regarded GIS modelling as their most rewarding activity, but warned it can be fraught with problems. For example, to create a successful model requires a level of expertise in GIS, the application area, and sometimes programming skills; and model outcomes are always mere approximations of reality and must be treated with caution. It is essential to have someone who is an expert in the application area interpret the data output. The following two quotes are typical of the survey responses:

> Modelling is what it [GIS] is all about. The problem is that to build a good model you have to be not only an expert in the subject matter you are attempting to model but also a programmer. It is very difficult to be both and almost impossible to excel in both.
>
> *Werner Runge, GIS Project Officer and Consultant*

> I use modelling occasionally to achieve certain desired responses or results from the data. But the biggest point to note about models is that

as the resident technical GIS staff, people will come to ask you for something—e.g. 'Can you combine vegetation with those soil complexy thingys and tell me all the woody vegetation on laterite soils?'—and it's up to you to turn that into a series of GIS commands.

Piers Higgs, GIS Consultant

A GIS user who is an expert in the application field, or who can communicate with the appropriate experts, and who understands the data, is an extremely valuable employee and the ideal GIS modeller.

Exercises: Weed vulnerability modelling for Jarrahlea

An analysis has shown that the areas where moisture is high, tree cover is low (incident sunlight is high), and soil is often disturbed are more prone to weed infestations than other areas. Figure 13.5a displays the weed infestation vulnerability map. A predictive model has been established, largely based on expert knowledge weightings of a number of factors influencing weed distribution. This model allows the GIS user to test two scenarios: one of clearing of trees in the Green paddock (Figure 13.5b), the other limiting movement along tracks (Figure 13.5c). The values, or levels of vulnerability, are indices rather than meaningful figures.

Q1. Examine Figure 13.5a. Can you suggest which layers have been used to construct this map?

Q2. How does Figure 13.5b suggest the clearing of trees will affect vulnerability to weed infestation on the property?

Q3. How does Figure 13.5c suggest preventing use of the tracks will affect vulnerability to weed infestation on the property?

Q4. Can you think of a method for validating the model?

Figure 13.5 Weed vulnerability modelling for Jarrahlea

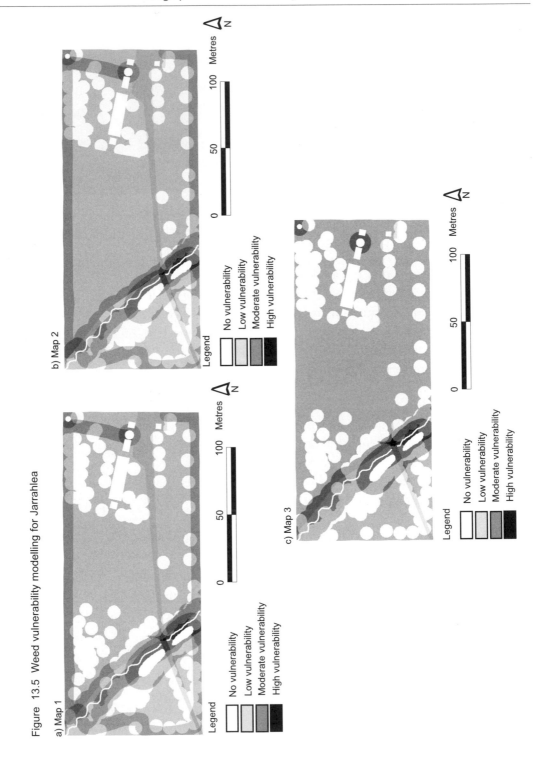

Where To from Here?

GIS directions

The direction GIS has taken in the past has been largely determined by the software developers and, more recently, the needs of the users.

GIS users demand more power and an increasingly user-friendly environment. The challenge is to meet these requirements with one software package. Software products need no longer be command-line driven; they may be windows-based and use buttons and menus. Yet some would argue that in order to allow full functionality, GIS packages still require command-line input (typing in commands using software-specific syntax).

The users ask for generalised algorithms (models) that are applicable in many circumstances, but the ability to tailor the model to any unique features of a locality or study is also needed. This requires the ability to alter a program, or to write in a programming language that can be interpreted by the software.

GIS users desire the software and machinery to perform faster, and to undertake more complex jobs, with more data, but to arrive at more accurate answers. Frustration stems from lengthy processing time, the inability to undertake complicated processing, and even small matters, such as numerical rounding errors. GIS users are beginning to question why GIS cannot perform some tasks that seem feasible in theory. Often, with the addition of other technologies such as artificial intelligence these needs are being satisfied.

With so many paradoxical requirements one cannot give a simple answer as to the future direction of GIS. The current situation is that software vendors are producing many modularised products in the hope of satisfying the needs of an ever-diversifying GIS user community.

GIS employment and the GIS work force

In the past, the number of potential GIS employees has often been too low to satisfy employer demands. The rapid introduction of GIS units or courses at school, college, and tertiary level will probably ensure a ready market of GIS users in the future. This, coupled with the growing number of application areas and increasing levels of acceptance in industry and government departments, suggests a bright future for GIS users.

GIS employment may be in one of two main areas—GIS in information technology, or GIS in application.

GIS users in information technology (IT) are the computer science GIS staff. They may be fluent in a number of programming languages, adept at constructing and managing databases, and capable of ensuring the GIS performs efficiently as a processing tool. It is important to note, however, that rarely will the IT-GIS user have training in the application field. Generally, to be most productive, they should work closely with an expert in the application field, such as a demographer or an ecologist, to avoid making incorrect decisions concerning the data processing and modelling.

GIS application users emerge from any number of backgrounds including geography, zoology, engineering and social science. They are able to use GIS as a means to an end. These people require and, at most teaching institutions, now receive basic training in compiling databases, in application programming, and other IT fundamentals. In essence, they are presented with a lightweight version of the computer science that will enable them to make informed decisions and to communicate effectively with IT-GIS staff. These GIS users often know exactly what they would like the computing system to do, but may not know the technical details, outside the GIS environment, needed to achieve the task.

In the past it has been common, because of budgetary constraints or misinformation, for many employers of GIS users to believe, or hope, that one person is capable of undertaking a number of tasks. These tasks may include:

- the management of the data (dealing with issues such as security, database design, and systems design and management);

- the management of projects (dealing with issues such as the best way to obtain data, time-frames, and budgets);
- undertaking the initial work (tasks such as buying, digitising, and attribute tagging data);
- conducting basic and high level analyses;
- writing and implementing models; and, on top of everything
- making nice maps.

Unfortunately, this mythical employee requires degrees in geography, information technology, management, mathematics, and cartography, as well as the area of specialisation.

As GIS databases have become recognised as valuable products, which often may be used to generate income, more attention has been focused on developing a GIS team of people from diverse backgrounds. This has proven to be, and continues to be, the best basis for an efficient and effective GIS department.

So, is it possible to be employable and valuable in the GIS community without high-level computing knowledge? With a good working knowledge of GIS concepts and software, you will be a valuable element in the GIS community. With a valid application base, you are also an indispensable element of the system itself.

The voice of the GIS user

In the survey, respondents were asked their ideas of where GIS in Australia was heading and to advise future GIS users.

Overwhelmingly, the respondents felt that GIS had influenced their career directions dramatically and was to become more important in many fields.

> I believe that in the future, most people working in natural resources management and environmental management will have some GIS skills, similar to the way most people today are familiar with word processing packages.
>
> *Narah Stuart, GIS Modeller*

> GIS has profoundly changed my [approach] as a botanist. It has provided me with a way of answering questions that I could previously only ponder.
>
> *Matthew Aylward, Part-time GIS Operator*

The most important advice given was to not dissociate the GIS processing and modelling from the real world. A good GIS user must understand the data and the processes involved in modelling. One day of field work may unravel many mysteries in the data that would never be solved sitting at a computer terminal.

> Get out from behind your [computer] desk once in a while and take a walk in the real world.
>
> *Nick Middleton, GIS Consultant*

Other respondents suggested ways in which the GIS community, as a whole, could improve. The main areas mentioned were the need for greater data-sharing between agencies and the necessity of encouraging one and all to keep high quality metadata and data dictionaries.

> I think the information exchange side of GIS is our biggest challenge. We need to have better avenues for exchanging data—products that allow easy format exchange, centralised data servers, etc. Far too much GIS grunt work is being duplicated because different organisations don't talk to each other.
>
> *Piers Higgs, GIS Consultant*

Where to now?

The practice of GIS is open to you. You now have a foundation for your GIS education, and you should progress to your chosen software package and more advanced readings. The example texts and reference papers provided (see References and further reading) are good places to gather more knowledge.

There are many software packages available. You would be well advised to speak to GIS users in the appropriate field of application in order to determine which packages are in common usage before enrolling in a software-specific practical course. You will, however, find transition between packages relatively easy with a good background in GIS concepts.

The Internet allows access to a number of good, free GIS software packages and will also enable you to track down data. Some data are free, although free data tend to be generalised or dated.

The future looks bright for students of GIS in terms of employment, new research application areas, and the release of more powerful, user-friendly software products. GIS users are equipped with a powerful and exciting research and analysis tool.

Appendix 1:
Answers to the Exercises

Chapter One: Exploring Jarrahlea

A1. There are eight layers shown on the map: paddocks, house, sheds, dams, creek, fruit trees, marri trees, and tea-trees.

A2. The area of the property is 24 513 square metres. How long did it take you to estimate this? The GIS gave the answer in, literally, a second!

A3. Two thirds (28 out of a total 42) of the tea-trees are within 10 metres of the creek. Again, the GIS was powerful enough to give this answer in seconds.

Questions 2 and 3 illustrate why GIS has become an efficient and effective tool in many workplaces. The ease of calculations using spatial and attribute data in GIS is phenomenal.

Chapter Two: Jarrahlea data recognition

A1. The point layers show marri trees and water bores. The line layers display the creek and the bridges. The polygon layers are water tanks and paddocks. Some symbols and thick lines can look like polygons (areas). The legend should help differentiate between these geographical primitives.

A2. Interestingly, at a different scale, the layers may differ in the manner of the data representation. For example, if the **map extent** was only one paddock this would enable the trees to be displayed as polygons of the tree crowns and the bores as polygons showing the bore housing structure. Similarly, the creek or the bridges could be drawn as polygons. If the user were to alter the scale to a regional scale map, the water tanks would be best represented as points.

A3. The creek has been converted from a vector line layer into a raster grid cell layer. Notice the way in which the spatial data has been altered as a result of the conversion. The creek is now very atypical or 'blocky' in appearance. Cells appear outside the original map extent (in the upper left-hand corner of the property).

    The water tank polygon layer was converted to a raster layer and then converted back into a smoothed vector polygon layer. The result is a polygon depiction of water tanks. The polygons do not appear blocky; however, distortion from the original product is evident.

A4. The most obvious answers are as follows:
    spatial data: 'centred approximately 95 metres south-west of the homestead'
    attribute data: 'Footrot Flat'
    topology: 'adjacent to'

Chapter Three: Capturing the Jarrahlea digital data

There are no correct or incorrect answers to these questions. The aim was simply to start the reader thinking about the alternative data input methods and the appropriateness of each method to a particular data layer. For the reader's interest, the data capture processes are described below:

A1. The fences layer was initially digitised (I) from the aerial photograph (B), as the fence lines did not appear on the topographic map. The scale of the aerial photograph was 1:25 000 and the property occupied only a small portion of the photo, making the identification of fences difficult. Field work (D) was undertaken to overcome any errors. The field data was incorporated through the use of the keyboard and mouse (IV). Field data were also used to edit the digitised data gathered from the aerial photograph.

A2. The tracks were very difficult to identify on the aerial photograph and required surveying in the field (D). Locations of tracks were drawn onto a base map of the property. This map was then digitised (I) using the digitising board.

A3. The gate locations were similarly captured from a field survey (D) of the property. A digitiser (I) was used to enter the data.

A4. Dam locations could be seen on the topographic map (C). The topographic map was purchased in digital format from the data custodian. The digital data was imported (III) into the GIS.

A5. The spa location was recorded accurately and precisely on a plan (A), which was drawn up prior to the spa's construction. This plan was digitised (I) into the GIS environment.

Chapter Four: When Jarrahlea is not as it should be

Spatial errors

1. One of the end points (nodes) of the line that was digitised as the boundary of the paddock has been entered incorrectly (towards the top right-hand side). This will affect the ability of the paddock to be displayed and analysed as a polygon layer. It will be impossible to add the name, or any attribute data, to the Paddock Attribute Table as a polygon label. Area will not be calculated.

2. There is an overshoot on the house polygon. This extra section of line work will not prevent the house being built as a polygon. Some GIS packages will not build the entire layer topology due to the dangle.

Attribute errors

Looking carefully at the attribute table reveals that the data entry phase has been quite careless.

3. Record 4 has a type without an initial capital letter, i.e. plum instead of Plum. If the user wished to select all plum trees for an analysis, it would be necessary to select records with types Plum or plum.

4. Record 5 has a typing error, i.e. Peac instead of Peach. Selection problems similar to those mentioned above would occur with Peach trees.

5. Record 6 also has a typing error, i.e. Cheery instead of Cherry. Selection problems similar to those mentioned above would occur with Cherry trees.

6. Record 7 has an incorrect height unit. This height should have been 0.95 metres. The height has been given incorrectly in centimetres. This will affect any calculations relating to the height attribute, such as the average tree height.

7. Record 9 has incorrectly used a character (letter) in a numeric attribute. Height is given as 2.1o instead of 2.10. This error will prevent any calculations being undertaken.

8. Record 13 has missing data for the tree type. It is useful to use a text string such as 'Missing' rather than leaving an entry blank. A blank entry can mean the attribute was unrecorded, unknown, or non-existent.

Topology errors

9. The paddocks layer will have topology built as a line layer rather than a polygon layer. This means that there will not be a polygon label point for the paddock, as discussed previously.

10. The house will have polygon topology. Most GIS software packages will inform the user that some of the spatial data in the house layer has not been used in constructing topology. This indicates that a dangle, or overshoot, is present.

Chapter Five: Coordinating Jarrahlea data

A1. Figure 5.3a has a grid of digitiser units. The map was placed on the digitiser just to the right of the digitiser grid origin. The original map was approximately 48 cm by 20 cm. This is unlikely to be screen units as it means the screen would be larger than 48 by 20 cm!

A2. The units shown on the grid in Figure 5.3b are distance units. These may be used if the coordinate system is not known; however, the actual distances between points are known. Working with a system such as this will allow overlay of the layers and simple calculations, such as area and length.

A3. One MGA94 unit is 1 m in distance. The first value in each coordinate pair is the **x-coordinate** and the second is the **y-coordinate**.

A4. The equal-area cylindrical projection has the greatest distortions towards the polar regions. Note the shape of Tasmania.

Chapter Six: Getting the Jarrahlea data out of your system

A1. The simplest and most effective product for illustrating spatial patterns is a map. The devices required may be a screen, a plotter, or printer for a hard-copy product.

A2. An indication of the relative areas can be given in a number of ways. A chart may be appropriate, such as a pie chart (Figure 6.3b), and a table (Figure 6.3d) may aid interpretation. The choice between the chart and the table will depend on the complexity of the data and the number of categories, as well as the preferred format of the data interpreter. If precise values are needed, a table may be more beneficial. A map can also be used to show the relative proportions; however, as the variability and number of classes increase, the ability of the user to arrive at an interpretation of relative proportions is lowered. The devices required may be a screen, or a plotter or printer, if a hard-copy is required. If the proportion figures are to be used in other software they may be produced in the form of a digital file.

A3. In order to satisfy such a loosely defined request for information, the GIS user would probably compile a report. The report may consist of maps, graphs, charts, tables, and statistics. The report may be submitted in a digital format or be printed as hard-copy.

Chapter Seven: About the data about the data

A1. Figure 7.5b is a data dictionary, which explains the codes used to map the distribution of exposed soil in the Jarrahlea paddocks. Figure 7.5c displays a metadata extract, providing details about the layer of exposed soil.

A2. The data displayed are updated each year using field survey techniques. The time of update is September, when the ground cover could be expected to be close to maximum. The data are highly variable and the level of soil exposure could possibly vary from month to month or day to day. The study at hand, however, investigates the changes from year to year. As the data capture will continue indefinitely, the 'Data Ending:' entry has been labelled 'ongoing'. The data from each year may be archived as unique layers, such as Exposed Soils 1998, Exposed Soils 1999; however, the current analysis layer will always be Exposed Soils, and will contain the most recent data.

Chapter Eight: Listing, displaying, querying, reclassifying, measuring, and reporting on Jarrahlea

A1. *What is the length of the creek?* This is an example of the use of a measurement tool. It is used to measure the length of a linear feature. *What is the area of the surface of the two dams?* This is another example of a measurement. This time the measurement concerns the area of two spatially separate features in one layer. *Which paddock is located 40 m due south of the large water tank?* This is a query issued using spatial and attribute data. The query identifies a polygon feature using an attribute label. *What is the feature the cursor is pointing towards?* Again, this is a query. The query is based solely on spatial location. The response is a listing of attribute data relating to the chosen spatial location. The map itself is an example of a display tool. The figure shows the paddock's spatial extents and names, dam locations, creek lines, bridges, water tanks, and bores.

A2. *Can we alter the marri and fruit tree display to show the spatial distribution of young, moderate and old aged trees?* This is an example of a reclassification. Two layers with similar attribute data have been displayed using the same scheme, in this case different sized circles. The layers are still unique and have not been combined. *Supply the details pertaining to marri trees.* This is an example of a listing tool. The attributes associated with the layer are listed, then the individual records are listed along with their attribute values. *Give summary statistics relevant to the vegetation in the Homestead Yard.* This is an example of a reporting tool. In this case the results are given as raw data, numbers of trees, and as percentages of the total tree population and the tree population in each unique layer. The figure has two examples of displays. The data in the displays are the same; however, the attribute labels used for creating the display have been varied.

Chapter Nine: Geoprocessing Jarrahlea

A1. The erase tool was used to create Figure 9.12a. The road was erased from the paddocks layer. The extract tool was used to create Figure 9.12b. The road was extracted from the paddocks layer. In this case, the extract and erase tools could be interchangeable. However, using extract for Figure 9.12b and erase for Figure 9.12a was most efficient.

A2. The built structures that have been buffered are the water tanks, the house, the spa and the sheds. The property boundary has only been buffered towards the inside of the linear feature of the fence line. If the fence line had been buffered to the outside as well, the new data layer would include a spatial extent larger than the original extent. Data outside the study area is not of interest.

A3. The paddocks which have been merged into larger polygons are Burrow, Fruit Grove, and Road, all of which have a sparse grass cover; and Footrot Flat and Bore Paddock, which both have moderate grass cover.

A4. The update process has updated both spatial polygon boundaries and attribute data for the areas defined spatially by the fire break (the buffered property boundary). After clearing, the percentage of ground showing exposed soil increases. All other areas remain unaffected.

A5. In terms of the type of data (point) and the spatial locations of the points, Figures 9.14b and 9.14c are exactly the same. Therefore, looking at the map alone will not enable the user to recognise that the overlay has taken place. The table next to the map, 'Attribute Table for Overlay Layer' shows the user that the attribute data for the paddocks have been added to the attribute data for the marri trees.

Chapter Ten: Layer upon layer of Jarrahlea data

A1. The chosen values have enabled the two grids to be added to give a unique resultant value for each possible combination of input values. Quite often the GIS user will need to recode cell data into arrays with difference ranges to achieve this type of result.

A2. The cells that are moist in summer and dry in winter indicate where the irrigation line runs across this small section of the property. These cells have a value of 21.

Chapter Eleven: Proximity analysis amongst the tea-trees and networking with Jarrahlea

A1. There are two linear features that might be attributed with network attributes, such as volumes of flow and direction. These are the creek and the tracks. The creek water flows from the south-east to the north-west. The volumes vary seasonally. A network would not be very useful for the data presented here, as there is only one section of stream. If the channel was branched, then an analysis may prove more useful.

The tracks might be viewed as paths for human movement. Weightings might be applied to sections of tracks based on the slope of the ground and the soil stability. Where there is more than one route for movement from A to B, an optimal track could be chosen based on the weighting. Again, as the number of unique sections of track is low, a network study would be an inefficient use of time and resources.

A2. The tea-trees appear to be clustering in space. They also appear to be in close proximity to the creek, possibly due to the influence of over-bank flow or simply due to moister soil in these lower regions of the property.

Chapter Twelve: The highs and the lows of Jarrahlea

A1. The creek can be seen as a white region, the palest area within the grid layer. The creek line is also the lowest-lying area on the property.

A2. The small triangles around the creek indicate that the creek area has a larger relative variability in elevation in comparison to other regions of the property. These areas also had more data points in order to represent the greater variability. The edges of triangular elements line up along the centre of the linear creek feature. There is another clustering of small triangles towards the eastern side of the property. These represent a steep slope leading to the house. The land was levelled for construction of the dwelling and this slope represents the material added to the hillside in order to create the level surface. The large triangles illustrate area where there is relatively low variability in elevation.

A3. The mesh has been made to appear three dimensional by raising and lowering nodes based on their elevation values. The vertical exaggeration is very low. To make the rises and falls more obvious the exaggeration could be increased.

A4. Aspect is difficult to display effectively as the values range from 0 degrees (north) through to 360 degrees (which is also north). A normal colour ramp is therefore unsuitable. Note how the colour scheme used in Figure 12.13a ranges from dark northerly aspects to lighter southerly aspects. Much of the property is south-facing.

A5. The very steep slopes are the dark regions. There appear to be steep slopes around the creek, next to flat creek banks. There is a steep region around the house and sheds, showing the edge of the area that was levelled for building construction. In the north and north-west of the property there is an extremely steep area. This is a natural rise and continues into the next property.

A6. The aspect and slope maps were created from the TIN. If you look closely, there is evidence of the triangular elements. The aspect map also appears to have been rasterised, suggesting that the data were converted into a grid layer either before or after aspect was derived from elevation values.

A7. The steepest section of the walk will occur at the location approximately 125 m along the x-axis.

A8. No, the observer cannot see the target in this visibility analysis. The rise of the land between the observer and the target hides the gate.

A9. The errors in the flow accumulation are occurring around the edges. Note the white areas in the north section, particularly in the north-east. There are areas that drain into these cells; however, they are in the adjoining property and therefore are not part of this data set. Towards the southern end of the property the flow accumulation looks high. Again, this is occurring because we have no information about the drainage beyond the property boundary. To avoid these errors, an entire water catchment should be used in the modelling.

The model does give an idea of relative levels of flow accumulation as the high flow accumulation values drain towards the creek and the dams, as would be expected.

Chapter Thirteen: Weed vulnerability modelling for Jarrahlea

A1. The layers used in the modelling include the creek, trees of all species, tracks, buildings (house, sheds, tanks, and spa), paddocks and dams.

A2. The clearing of trees has broadened the area susceptible to weed population growth in Green Paddock. One factor associated with likelihood of infestation was the amount of impinging sunlight. The removal of trees increased both incident sunlight and the vulnerability levels. Note that in this output the variability in weed infestation vulnerability of the Green Paddock has been dramatically lowered.

A3. Closing tracks to use will lower the vulnerability of these areas to weed infestation. As the populations are already established here, the owner will need to remove the current population and try to revegetate the land with native vegetation.

A4. The model is very difficult to validate, as the index is vulnerability to infestation rather than a measure of actual, real weed distribution. The user could map current weed locations and determine whether the modelled highly vulnerable areas have relatively dense weed populations and the modelled low vulnerability regions have relatively sparse or no weed populations. These results would not be conclusive, as factors such as the mechanism of weed dispersal, current population data, and management factors are not included in the model. An area modelled as highly vulnerable may have no weeds apparent in the real world, simply because the weed seed or spores have not reached these locations and therefore have had no opportunity to establish.

Any model is an approximation of reality, and should be interpreted with this in mind.

Appendix 2: Australian GIS Stories

GIS in action

This appendix contains case studies of GIS applications in Australia. The studies were chosen to give the reader a diverse and interesting set of example implementations by GIS users, involving tools discussed previously in the text. The first is an academic GIS project with the potential to become a very important application. The second case study uses GIS and GPS to help serve government and community groups. The third paper comments on methods available to remove or overcome spatial errors, and the final section contains information regarding an organisation helping others access and make the best use of data.

The author, Julie Delaney, would like to thank David Smith, Toni Furlonge, Tim Molloy, Carl Bennet, Andrew Burke, and Robin Piesse for permission to reproduce the papers given in this appendix.

Case Study One: Mapping the bushfire danger of Hobart

Author: David Smith, Centre for Spatial Information Science, School of Geography and Environmental Studies, University of Tasmania

The January 1998 fire in Hobart's southern bushland was a reminder of the city's vulnerability to this familiar phenomenon. Seven houses were lost and an area of approximately 3100 ha was burnt.

Bushfires are an ever-present threat to life and property in the City of Hobart during the warmer drier months. There are a number of factors that contribute to this threat:

- 56% (4320 ha) of this local government area is covered by bushland, so there is a significant presence of fuel for bushfires;

- Hobart is located in south-eastern Tasmania which experiences the most extreme fire weather in the state; and
- since the early 1970s there has been constant population growth in the interface region between Hobart's suburbs and the adjacent bushland.

Since March 1997 a definition of *fire hazard* has been used by local government authorities in Tasmania to classify the bushfire threat or danger at any location. Essentially, the definition classifies sites as hazardous (bushfire prone) if they are within, or in close proximity to, substantial areas of vegetation. Bushfire prone land with a slope of up to 15° is rated as Moderate hazard. If the slope exceeds 15° the hazard rating is High. Land not designated bushfire prone is given a Low bushfire hazard status. If a location is bushfire prone and is the subject of a planning application, the local council would consider the slope at the location and other relevant factors, such as access to the site and water supply at the site, before deciding whether to allow the development to proceed.

The concept of *fire danger* incorporates potential fire behaviour, the probability of ignition, the suppression response of fire management organisations, and the values (both natural and constructed) at risk. In this project, potential fire behaviour was mapped as a representation of fire danger.

The aim of this GIS project was to:

- apply the Tasmanian definition of fire hazard to map the fire hazard of the City of Hobart;
- apply a model of potential fire behaviour to map the fire danger of the City of Hobart; and
- use the mapping to determine the effectiveness of the fire hazard definition in predicting the modelled potential fire behaviour.

Potential fire behaviour was modelled in terms of potential fire intensity based on procedures used by the Department of Conservation and Land Management in Western Australia, Planning South Australia and the Country Fire Authority of Victoria. Fire intensity is a useful measure of fire behaviour because it indicates:

- whether control of the fire is possible; and
- the damage that may result from the fire.

Both fire hazard and potential fire behaviour are determined by factors that show spatial variation. GIS with raster modelling capabilities allow modelling and analysis to be efficiently performed with spatially varying factors. For this reason, GIS are widely used to support fire management in Australia and they provided a suitable means of carrying out this investigation.

The slope layer required for the mapping was derived from a digital elevation model (DEM) of the area. This raster DEM was interpolated from 10 metre interval contours. Polygons representing reservoirs and the River Derwent were also used as inputs to the interpolation process, giving rise to flat interpolation at these features rather than depressions. The shoreline of the River Derwent also provided a zero metre contour.

The cell size of the raster layer should ideally be as small as possible, in order to optimise the modelling of the terrain. However, the resolution of a DEM is limited by the accuracy of the data from which it was derived. The Department of Primary Industries, Water and Environment, from which the contour data originated, states a horizontal positional accuracy of within 12.5 m 90% of the time. Hence, a cell size of 15 m was chosen. This cell size was used for all the raster layers derived in this project.

The McArthur model of forest fire behaviour was used in the calculation of potential fire intensity. This empirical model has been widely used for fire management in southern Australia since the 1960s. The required inputs for the model include fine fuel loads, meteorological variables (air temperature, relative humidity, wind speed, and a drought factor reflecting the dryness of fine fuel), and slope.

Vegetation communities for which there was fuel data (73% of the bushland) were reclassified as fuel types. Fuel accumulation curves were then used to generate fuel load layers, representing predicted fuel loads 5, 10, 15, 20, and 30 years after the last fire.

A forest fire danger index (FFDI) summarises the fire weather determined by the meteorological variables. In this project an FFDI representing a median high fire danger day in Hobart was used, based on observations made over eight fire seasons at the Bureau of Meteorology's Hobart measurement site. This FFDI was applied to the entire study area.

The chosen FFDI, the slope layer, and the fuel load layers were then used to calculate potential fire intensity layers for each of the five fuel accumulation periods. The worst case, in terms of the wind's effect on a fire, was modelled by assuming that the wind was directed upslope at each location.

The fire hazard map was then overlaid with each of the five potential fire intensity maps for comparison purposes. It was assumed that the fire hazard definition predicted potential fire behaviour correctly if:

- a location at which there was potential for a fire that could be controlled was classified as Moderate hazard; or
- a location at which there was potential for an uncontrollable fire was classified as High hazard.

A critical skill index (CSI) was used to summarise the degree of critical misclassification (a location at which there was potential for an uncontrollable fire being classified as Moderate hazard). The CSI could vary between 0 and 1, and would ideally be zero.

The fire hazard definition predicted correctly at least 75% of the time for four of the five fuel accumulation periods used. The optimum success rate of 89% occurred for 15 years of fuel accumulation. In the least successful case (5 years of fuel accumulation) all the misclassifications were non-critical. The CSI was very good (≤ 0.11) for the first 15 years of fuel accumulation. It also indicated reasonable performance (≤ 0.27) for the following 15 years. In assessing these findings it is important to consider that fuel accumulation beyond the first decade is generally relevant to just the wet forest areas, since the dry forest areas are likely to experience a fire during this time period.

The accuracy of the DEM was estimated using the heights attributed to 188 survey control point locations within the City of Hobart. The standard deviation of 1.13 m provided a means of estimating slope uncertainty. The fire hazard definition's success rate and the CSI showed some sensitivity to the estimated uncertainty in slope, but the overall conclusions reached were similar. The results showed low sensitivity to the estimated uncertainty in the locations of the vegetation boundaries.

The findings of this project would give those responsible for land use planning in Tasmania confidence in the fire hazard definition that they are using. The results suggest that the definition is reasonably effective in either predicting potential fire behaviour, or at least not misclassifying this behaviour in a way that has critical implications.

I would like to acknowledge the advice and support provided by my supervisors, Dr Eleanor Bruce and Associate Professor Manuel Nunez, and Mark Chladil of the Tasmania Fire Service.

Case Study Two: Geographic Information Systems Support in Implementation and Monitoring in the Goulburn Broken Dryland Salinity Management Plan

Author: Toni Furlonge, GIS and Remote Sensing Unit, Catchment and Agricultural Services, Benalla, Victorian Department of Natural Resources and Environment

This project is based in Benalla, Victoria, and is supporting dryland management plans across the Goulburn Broken Catchment Dryland (GBC). It has a large emphasis on working with landcare and community groups, along with other agencies involved with dryland salinity management plans. The GIS unit developed from this project has expanded to work across all natural resource issues in the north-east region of Victoria.

Our awareness and knowledge of the environment in which we live is leading us to believe that the 'problems' that are impacting on the 'resources' are symptoms of an ecosystem under stress. This project is about helping to define the 'problems' and the 'resources'; recording the efforts in combating the 'problems'; assisting the 'resources' to recover; and monitoring the effects of our efforts.

GIS support in implementation and monitoring in the GBC is a 3-year project launched by the Department of Natural Resources and Environment (DNRE) in April 1996.

The GBC forms part of the Murray Darling Basin and is responsible for 11% of the water flowing into the Murray Darling Basin (Goulburn Broken Catchment and Land Protection Board, 1985).

The GBC covers 2.4 million hectares, which constitutes 17% of Victoria. The catchment land uses include water (47 000 ha); state forest (414 300 ha); parks (95 800 ha); urban areas (3600 ha); and alpine resorts (4200 ha).

Salt loads flowing into the Murray River are increasing and it is predicted that they will continue to increase substantially in the future. This is a cause for concern as the land and water systems will continue to degrade in the dryland and irrigation regions of the catchment. As a result it is essential to record, monitor, and evaluate current processes, management techniques, and qualities of the environment. Such a database would assist with the management of catchment health, provide data to assist with the best management practices, and also provide data for modelling future scenarios.

The databases established will support landcare community groups, DNRE, local government, water authorities, environmental consultants, and catchment management authorities in the development and implementation of Dryland Salinity Management Plans, Whole Farm Plans, and Local Area Plans.

Existing data capture and storage methods available required updating. A GIS was established with a number of datasets, databases, and other information products. The GIS employees also fostered an awareness of GIS and GPS technologies. The GIS software was based both in Melbourne and locally in Benalla.

Once catchment members openly embraced GIS, the introduction of GPS commenced. The GIS team uses three GPS **Dataloggers** and their own community base station. All data captured by GPS is differentially corrected and used in the GIS software.

The GIS team has adopted the roles of adviser, trainer, and supporter to landcare community groups, and agencies, enabling them to do the data capture using methods and standards developed and set by the GIS DNRE CAS Benalla. This enables the GIS team to specialise in the further development of the databases and in the development of information products in support of the GBC. All products are then used by the landcare community groups and agencies in their recording, monitoring, and evaluations of on-ground work programs and management planning.

The GIS team has trained and supported DNRE officers, community groups, water authorities, natural resource consultants, and local government across the north-east region. This has been in the form of:

- developing site specific databases;
- providing assistance and equipment for data collection, storage, and analysis;
- developing information products such as maps and spread-sheets;
- providing a service for GIS and GPS giving advice and training;
- developing local area and catchment management plans; and
- assisting with the development of Salinity Management and Whole Farm Plans.

A comprehensive GIS database has been developed; some of the layers are:

- 1:25 000, 1:100 000 and smaller scale, road, hydrology, contours, and **cadastre** data
- Satellite imagery
- Wetlands locations
- On-ground work programs
- Community groups action areas
- Salinity extent
- Saltwatch, Waterwatch monitoring sites
- Soil types
- Sub-catchment locations
- Vegetation types and layer
- Pest plant and animal distribution
- Site specific datasets
- Local Area Plans
- Environmental threats and values

GPS is used by landcare community groups and agencies to develop GIS datasets for the following:

- Local Area Planning. Community groups are provided with base maps to record environmental threats and values. From this management and work plans are developed.
- Land Protection Incentives. Grants are given to landholders for on-ground works, which will lead to improvement in catchment health. GPS is used for site assessment and mapping. Site plans and catchment overviews are produced using the GIS.
- Pest Management. GPS is used for site assessment and mapping of pest plants, and animals. Site plans and catchment overviews

are produced using the GIS. The GPS data is exported to databases in the GIS.

- Creek Assessment. Landcare groups use the GPS to document threats (erosion, stock access, and pests) and values (vegetation) along creeklines.
- Whole Farm Plans. Individual landholders request that property and paddock boundaries be mapped. Existing paddock areas are determined. New paddocks are designed based upon size and terrain, using GIS.
- Farm Forestry. Involves site assessment and mapping of plantations being established across north-east Victoria and the production of site plans and regional overviews.
- Base maps are produced for the development of other GIS data.

The GIS database includes imagery from air photos to satellite imagery.

Work has been done with community landcare groups developing methods to use satellite imagery (Landsat TM) to map perennial pastures, bracken and Paterson's curse. This is done using the local knowledge of seasons and weather conditions for the classifications and for ground truthing classified imagery. This satellite image product is then useful for the landcare group as it contains clear, interpreted information.

The GIS Department of Natural Resources and Environment Catchment and Agricultural Services, Benalla, will continue to provide GIS, Remote Sensing and GPS services to the north-east region of Victoria for:

- contract mapping, catchment, and local area plan development;
- cost benefit and evaluation analysis;
- community, departmental, and agency strategic planning; and
- evaluating change.

GIS and GPS equipment and expert staff are available to develop datasets, catchment, and local area plans. This is supported with a community base station for metre accuracy GPS mapping, a comprehensive GIS database and related databases with supporting documentation, and expert staff.

The GIS group will continue to develop local area, sub-catchment, salinity, and whole farm plans, and continue mapping environmental and natural resource threats and values for agencies and landcare community groups.

Access to GIS datasets and technologies will continue to increase with packages being customised to suit specific sites and office requirements. Support and training will continue to be provided across agencies and the community.

Case Study Three: Overcoming Inaccuracies in the NSW Digital Cadastral Data Base

Authors: Tim Molloy and Carl Bennet. Tim Molloy is the GIS Coordinator at Hastings Council, and Carl Bennet is the GIS technician at Hastings Council.

The Hastings Council is located on the Mid North Coast of NSW. It has had to overcome inaccuracy problems with the digital cadastral data base (DCDB) by upgrading the DCDB using software (ACRES) written by Bob Clatworthy.

The ACRES software allows rebuilding cadastre using bearing and distance entry from subdivision plans. When this survey accurate data is attached to the state survey control network, a survey accurate coordinated cadastre is able to be built.

The Hastings Council's current plans for the use of the ACRES software is to completely rebuild the cadastre for all urban areas within the Council area, focusing initially on urban growth areas. This enables the digital entry of survey accurate asset data for new subdivisions (water, sewer, roadworks, etc.) submitted as Work as Executed Plans to be dropped straight into Council's GIS. This will happen while maintaining the survey accuracy of the data and eliminating the time consuming process of rubber sheeting (or warping) the survey accurate data to fit the (by comparison) inaccurate DCDB data. Council has found that in urban areas the accuracy of the DCDB was generally in the order of ±4 m while in rural areas the accuracy was in the order of ±30 m.

The Hastings Council had previously been entering subdivision data digitally; however, in order to adjust adjacent subdivision plans to fit together each individual traverse entered required an adjustment. Using the ACRES software proved to be a superior process to the old method both in terms of speed and accuracy as it has auto-mated the whole adjustment process.

The ACRES software also allows upgrading of the DCDB to a desired accuracy using survey control on a number of points within the cadastre. This 'upgrade' feature was Council's initial use of the software. Survey accurate sewer data including coordinated property boundaries had been captured by Council's Waste Water section. In order to incorporate this survey accurate data into the GIS, the accu-racy of the DCDB needed to be improved. This was achieved by using the coordinated property boundary ties as control for the ACRES software upgrade.

One problem encountered in the upgrading process was maintain-ing the relative spatial accuracy of the upgraded DCDB, particularly where digitising of the original cadastre was not uniform. Features of the ACRES software allow different weighting of polygons to ensure they maintain their relative shape. In some instances gross errors were found such as two adjacent subdivisions being digitised out of relative alignment by up to 20 m. In these instances with relative spatial accuracy being almost non existent, a more practical approach was to completely rebuild the cadastre in these areas.

Another major time saving feature of the ACRES software is the Utility Transformation Manager (UTM). This feature allows trans-forming of other GIS layers to maintain relativity with upgraded cadastre. UTM uses the start and finish coordinates of the upgraded cadastre to control the upgrading of other layers. A special technique had to be developed by Hastings Council in order to generate start and finish coordinates where the cadastre was rebuilt as these coor-dinates were not held in the ACRES software.

The upgrading program has enabled Hastings Council to:

- enter asset data into the GIS with greater speed and accuracy;
- provide upgraded cadastral accuracy; and
- upgrade the accuracy of other data layers in the GIS.

Case Study Four: Western Australian Land Information System

Authors: Andrew Burke and Robin Piesse, WALIS

Established in 1981, the Western Australian Land Information System (WALIS) is the longest standing LIS/GIS cooperative arrangement in Australia and perhaps in the world. While WALIS has been through a few of the restructuring exercises that are an inevitable part of the public sector environment, the intrinsic philosophy and administrative structure has remained remarkably consistent.

WALIS now has 26 member agencies representing all spheres of government; it has an active industry advisory body and an attentive Executive Policy Committee comprising the Chief Executive Officers of the WALIS agencies.

WALIS Forum, the annual conference, attracts over 400 people, WALIS News reaches over a 1000. There are WALIS policies in place for custodianship, pricing and marketing, metadata collection, and standard licensing agreements, and all member agencies now sign a Memorandum of Understanding that specifies that they adhere to the intent of these policies.

But how did it all begin and what benefits does WALIS confer on the state of Western Australia?

Early Days

In the mid-1970s there were problems with the management of WA land information. Boom times had led to enormous increases in the volume of transactions in tenure and cadastral information and manual retrieval processes were cumbersome. There was a lack of interdepartmental communication that encouraged duplication.

In 1978, WA Government Computing Policy Committee commissioned a management study by PA Consultants to complete: 'An analysis of requirements and determination of a framework for a comprehensive **land information system** to meet the administrative needs of Western Australia'.

Significant duplication and allocation of resources in land information management were identified by the study that left plenty of scope for improvement.

The concept of a land information system (LIS) described in the ensuing report recommended the development of sub-systems to manage each particular type of land information. A common geographic reference system would then allow parcels of information from each sub-system to be related to unique geographical locations. In this concept the data is the greatest influence on system development, not the functions which may be performed on that data.

At this early stage WALIS enjoyed senior management and government support and the emphasis was on establishing business cases for further action, rather than assuming the application of technology alone can solve problems.

The initial emphasis was on transferring very large amounts of legal and graphic cadastral information into digital format. The aim was to integrate these systems to develop a property register for all Crown and freehold land.

WALIS was then to expand to include a larger range of land information. From late 1982, attributes such as street addresses, land use, statutory zoning, valuation records, and administrative boundaries such as local government, census districts, localities, and postcodes were to be progressively incorporated into WALIS.

In 1982 the structure of WALIS emphasised participation from government agencies, industry, and the community (Figure A2.1).

LISAPC—Land Information System Administrative Policy Committee

LISEMC—Land Information System Executive Management Committee

The Special Interest Groups (SIGs) represented land information 'sectors' such as rural, urban, administrative, education, utilities, and technical interests.

Unfortunately the role of the Land Information System Support Centre (LISSC) (and Land Information System Advisory Committee (LISAC)) had gone far beyond the coordination role initially envisaged, to that of a controller, and was seen as an enclave of technical and political power. This was particularly so after 1984 when

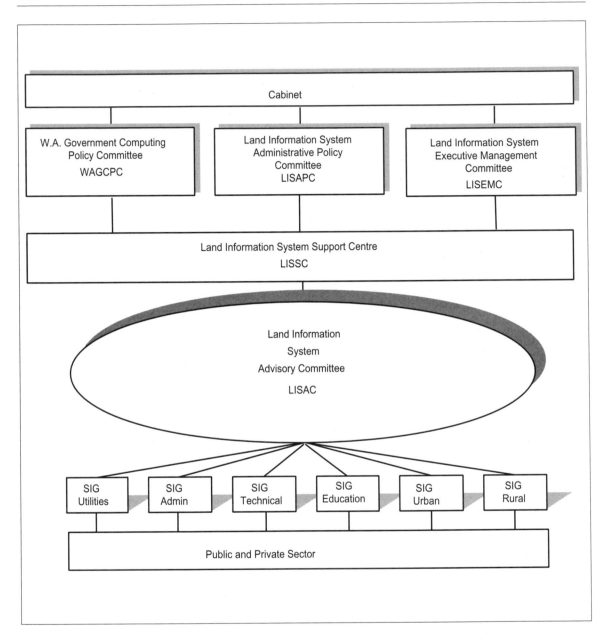

Figure A2.1 WALIS corporate structure

LISSC became part of the Department of Computing and Information Technology (DOCIT). WALIS was seen as a creature of LISSC, not as a creature of the participating agencies. This concentration of power was seen as untenable by many senior managers and was a major factor contributing to pressure for change.

Times of Change

The changes came with the formation of the Department of Land Administration (DOLA) in 1986. LISSC was transferred to DOLA and divided into two halves, one to address cross government coordination, that is WALIS, and the other to concentrate on issues internal to DOLA.

At this time several issues faced WALIS and these were articulated in the Strategic Development Plan, released in September 1987.

> How to physically develop an integrated system with emphasis on cross Government and efficiency gains yet allow agency specific developments to go ahead? To date, department-specific sub-system development had dominated WALIS, whilst issues of integration and access to corporate data for all the WALIS community had taken a back seat. Unfortunately, many of the recommendations received little support.

Some specific problems hindering the development of WALIS were seen as:

- slow cadastral data capture;
- computer overload—the corporate computer processing and storage capacities were 'hopelessly overloaded';
- lack of wide area network—even though DOCIT submitted a State Communications Strategy to the Government for all Government communications, no decision had been made. Without such a wide area network, on-line access was not possible.

Overall the organisational and technical difficulties in providing on-line access were consistently under-estimated. Not only with regard to the communications infrastructure or the systems development needed (in custodian systems and user systems), but also the data management regimes required throughout WALIS agencies to support on-line access.

Further changes came with a Cabinet Submission prepared with the objective of integrating and improving access to the State's land information. An Integrated Land Information Program Taskforce was established. In June 1990, Cabinet approved a number of recommendations, some of which follow:

- elevation of WALIS to full program status with DOLA and make Minister of Lands the accountable Minister;
- establishment of a State Land Information Capture Program; and
- allocation of custodial responsibilities to WALIS member agencies.

The aim of the Integrated Land Information Program was to 'integrate' the *system* rather than the *information*, with a series of distributed agency systems housing data, accessed through an efficient communication network. To achieve this, two priorities were to:

- develop adequate communication and networking strategies; and
- develop an appropriate policy and method of storing, maintaining, and disseminating corporate data.

Custodianship: Agencies formally accepted custodianship for nearly 100 datasets and established capture and maintenance programs for many of them. A comprehensive custodianship policy was developed in 1994.

Land Information Directory (LID): an on-line directory service providing metadata on WALIS data.

Networking Improvements: Adoption of **TCP/IP** communication **protocol** and NFS file server software provided full and transparent access between the **Ethernet** system and the System Network Architecture system used in WALIS agencies.

Upgrade of LIA: The Land Information Access (LIA) system was upgraded to provide a more integrated, accessible and accurate system.

Spatial Cadastral Upgrade: The most notable improvement was the conversion of the SCDB to a clean, complex topology able to support survey accuracy. Cadastral data was now to be collected at survey accuracy.

Data Pricing: Cabinet approved data marketing and common pricing policies for WALIS data in 1992.

In mid-1997 access to the Directory was improved with the development of the Interragator CD-ROM. Interragator supported interactive spatial searching for metadata as well as text-based searches.

A combination of studies conducted by ANZLIC and WALIS at this time showed a benefit cost ratio for Government and agencies participated in WALIS of 6:1.

Current State of Play

A major emphasis for WALIS today is improved access to land information. A distributed network approach, where users have access to a coordinated network of data directories, map browsers, and data warehouses has been adopted.

Improving access is a staged process, and the initial focus is improving access among WALIS government agencies. An updated Interragator CD-ROM has been released and Interragator On-Line is available through the WALIS website and linked to national directory sites. A prototype has been developed to provide a data warehouse of fundamental land information datasets to WALIS agencies.

The WA State node of the Australian Coastal Atlas has also been established by WALIS. The Atlas is an electronic atlas drawing together data holdings of the coastal zone from the Commonwealth and the States. Users can prepare maps and link to metadata records of coastal and marine datasets.

Working together WALIS agencies have funded a major project to validate and increase the layer of the Property Street Address database at DOLA. Property street address is a dataset fundamental to the operations of several government agencies and is of increasing importance to emergency service management.

Conclusion

Any information on WALIS today should not neglect the fact that there have been many world class LIS developments in WALIS agencies and a number of improvements to inter-agency information flow have been implemented.

WALIS has been successful in resolving many data management issues that will underpin future access mechanisms. For instance, custodianship, pricing, marketing, metadata, and data licensing.

The cross Government improvements delivering efficiencies promised when WALIS started still present a challenge. Improved telecommunications and software developments will only go part of the way to delivering whole of government solutions, the secret to success within WALIS is for everyone to work together.

To get the latest information on WALIS, visit the website: http://www.walis.wa.gov.au.

Glossary

accuracy In this text accuracy is used to refer to the shift between the location of the GIS data and the actual location in real space. For example, an accuracy of ±10 metres indicates that a GIS feature may be up to 10 metres from the true coordinate, in any direction. Attribute accuracy refers to the likelihood that the attribute label is correct for any spatial location.

attribute data Attribute data is non-spatial GIS data that is related to a spatial location. For example, attribute data may be a soil type or a landholder's name.

attribute definition Attribute definitions describe the format of the stored data. Attribute data may be real numbers, with 5 decimal places, in a column width of 10 spaces. A text string definition may be a 6-spaced text entry using an initial capital letter, a full stop, and lower case letters, such as 'E.cype'.

automated digitising Automated digitising is the term used for scanning information into digital format.

Boolean operators Boolean operators are used to make logical queries of two or more data layers. Example operators are AND (condition/s true in all inputs) and OR (condition/s true in at least one input).

buffering Buffering is a geoprocessing tool that creates a new polygon layer. The new layer displays all areas within a user-defined distance of features in another layer.

cadastre or cadastral This is information or data related to land parcels. Generally this will be spatial property boundaries and attributes such as land-owners and land values.

cartographic modelling Cartographic modelling in GIS is the combination of the techniques and operations, in an ordered manner, acting on data, to simulate a spatial decision-making process.

cartography Cartography involves combining science, drafting skills and artistic endeavour to make an effective map product.

central processing unit (CPU) The CPU is the part of the computer that actually undertakes all computations—the 'brain' of the machine.

command A command is an instruction to the software or computer program. An example command may be 'clear the screen' or 'draw the map'.

compact disk/disc (CD) or compact disk/disc read-only memory (CD-ROM) A CD or CD-ROM is a storage medium that is read by laser techniques. Generally, a CD is an audio disk and a CD-ROM is a computer-readable compact disk. CD-ROMs are ideal for archiving old data.

computer cartography Computer cartography is a science and an art dedicated to making effective maps using a computer.

computing platform A computing platform describes the type of computing environment, including the hardware and/or the operating system (a control program coordinating the computer).

continuous data Continuous data are attributes that exhibit smooth and continual changes across space. This means that any location in the spatial data will have an attribute value. Examples of continuous data include precipitation, elevation, and soil moisture. Continuous data are typically stored as raster grid cell data.

control point A control point (or marker) is a location that can be identified in unregistered data (say in digitiser units), and in registered data (from hard-copy or another digital source). This means that for a given location the user, or the GIS, must know the real world coordinates in a user-specified registration and projection, as well as the GIS-assigned digitiser or screen units.

data In this text, data is used to describe a known, which is used as a basis for deriving information. You may find digital data in GIS format on a disk (a hard or floppy disk), or in another format, such as a spreadsheet or database software program.

data analysis Data analysis is the process of interpreting data. This may range from simple exploratory data analysis (which involves simply looking at the data and describing what you see) to complex analysis such as looking for clustering patterns in the data.

data dictionary A data dictionary provides a key to codes and schemes used in a layer or a database. For example, a soils map may display soil types 1, 2, and 3. The data dictionary will describe that type 1 is sandy loam, type 2 is gravel, and type 3 is heavy clay.

data display This term refers to the appearance of data on the computer screen. Data may be viewed in any form, although the most common format is a map.

data input Data input involves bringing data into the GIS. This may involve importing digital data from other software, or capturing non-digital information into digital data format through a process such as digitising.

data management Data management refers to the way in which the system looks after the data, the way in which the user is able to access the data, and any security or integrity measures taken to minimise data corruption or deletion.

data manipulation Data manipulation is the way in which the data can be altered to suit the users' viewing needs.

data modelling In GIS this term refers to the way in which data can be used to predict or explain a spatially based process. A model may be a simple collection of any basic GIS functionality, such as a buffer process followed by an overlay, or it may be a highly developed, detailed approximation of reality, such as a greenhouse gas depletion model.

data output Data output (noun) is the product that results from processing. Data output (verb) is the process of moving digital information out of the software package into another environment. This may involve printing data out as a table or a map or retaining digital format as a file for import into another software package.

data retrieval Data retrieval is (1) the process whereby the user requests the data to be presented in a user-defined manner or a manner native to the GIS, and (2) the action being undertaken by the software. The way in which GIS brings stored data back to the user will depend on the database management system (DBMS) tools for data retrieval.

data set Data set is often used as a synonym for database, as it refers to a collection of related data. Strictly speaking, a data set implies a completeness that the database does not promise.

data storage In GIS data storage refers to the manner in which the data are kept in the machine. Data storage, therefore, is reliant on the tools offered by the database management system. This may be, for example, a compressed format, as an integral part of the software, or in a format that enables multiple software packages to use the data.

data structure The data structure refers to the manner in which GIS allows storage and display of spatial information. The most common structures are vector and raster.

database A database is a collection of organised, related data. Data may be related spatially, through like attributes, by data structure, or by any other logical means.

database management system (DBMS) A DBMS is a software tool that manages the database. It ensures integrity, consistency, and security inside a database.

datalogger A datalogger is a device that gathers and stores data, usually in the field. Global Positioning Systems are often dataloggers for a GIS database.

digital elevation model (DEM) A DEM is a representation of three-dimensional space. DEMs are used for creating a surface of elevation data showing the rises and falls across a landform. It is really a 2.5-dimensional model, as the elevation is stored as attribute data rather than a spatial element.

digital format Digital format refers to data stored within a computer in discrete units.

digital terrain model (DTM) A DTM is a surface made from a DEM. The elevation data may be used to create DTMs, such as aspect and slope.

digitiser A digitiser is a hardware device that allows line-work or points on a paper map to be traced into digital format. Manual digitisers include a digitising board, which may range from a small hand-held pad to a table that resembles a whiteboard in size. The tracing device, a puck, may resemble either a pen or a mouse. Automated digitisers are referred to in this text as scanners.

digitiser unit A digitiser unit is a measurement unit based on the wire mesh spacing within the digitiser. It is an artificial coordinate system (or georeferencing system or registration system) used to establish the correct position of features in relation to one another on the digitising board, and on the computer screen.

digitising Digitising can mean simply converting any information into digital data. More specifically the verb to digitise can relate to using a digitising device (see digitiser).

discrete data Discrete data have attributes that can be divided into self-contained, homogeneous areas. Variation of the attribute within the area (or polygon) is regarded as non-existent. Examples of discrete data include land use, lakes, and the states of Australia. Discrete data are typically stored as vector data.

disk drive A disk drive is a hardware peripheral device that reads from, and stores to, a disk.

drainage divide A drainage divide is a line that separates the areas of adjoining catchments. Drainage divide is synonymous with watershed.

drainage model A drainage model is, broadly speaking, any model related to flows across the landscape. Usually a drainage model employs a DEM as a primary data source. Examples include flow accumulation, topographic position, and drainage divides.

erasing (as a geoprocessing tool) This geoprocessing tool creates a new layer based on data in an existing, more complex layer, by erasing (deleting) unwanted data.

ethernet An ethernet is a local area network of computer cables (ethernet cables) used for sending data between machines.

explicative modelling These are models that attempt to understand (explain) relationships between data.

extracting (as a geoprocessing tool) This geoprocessing tool creates a new layer based on data in an existing layer, by extracting only wanted, user-defined data.

flow accumulation Flow accumulation is the cumulative sum of water flowing across a surface. Flow accumulation model outputs are grid cell layers in which each cell has a number representing the total number of cells that drain through that cell.

fly-through A fly-through is a journey created by the ability to literally 'fly' the user though a landscape using a series of images created from a DEM. It involves setting a start and end point, a height above the surface of the earth, and defining the path of travel.

geographical data Geographical data are spatially registered within a coordinate system and a projection system and possess associated attribute data.

Geographical Information System/s (GIS) GIS are computer-based software systems that allow input, output, analysis, and manipulation of spatial and related aspatial data.

geoprocessing tools Geoprocessing tools are part of the GIS toolbox. These tools are designed specifically for altering (processing) pre-existing geographical data to create new layers. Examples include the merge and buffer tools.

georeferencing Georeferencing is a process that locates data in real, geographic space using a coordinate system. Synonymous with registering.

GIS analyst A GIS analyst is a person who uses GIS for the purpose of modelling or analysis.

GIS user A GIS user is a person who uses GIS as an application tool. An example may be a planner using GIS for a site-selection task. The GIS user is not necessarily a computing expert.

Global Positioning System (GPS) GPS is a system that enables a position to be determined based on the relative location of satellites. Often GPS units are hand-held devices that can be used in fieldwork and can also be used to feed data directly into a GIS. A GPS primarily captures an x-coordinate and a y-coordinate; however, speed and elevation can also be captured.

grid algebra Grid algebra is a form of analysis of spatial data that is designed specifically for cell-based data or grid cell layers. Grid algebra involves manipulating data between different grids to create a new grid layer. Examples include addition and subtraction, which are also grid cell layer overlay processes.

grid cell A grid cell is one element of a grid layer and is characterised by one attribute value. Ordered collections of grid cells of identical size and shape compose a grid layer.

grid cell layer or grid layer A grid layer is a layer of data that is composed of grid cells. The cells are regular in size and area, and are usually square. Each cell contains a value of the surface attribute. Grid layers are ideal for the representation and modelling of continuous data.

grid DEM A grid DEM is a surface of elevation, usually interpolated from a point or line layer. Each square cell contains elevation data.

hard-copy A hard-copy refers to a copy of data that can be literally held in the hand. A hard-copy map is printed on paper rather than displayed on a screen.

hardware Hardware is the physical computing system of a GIS. Hardware consists of equipment that may actually be touched, and includes computers, screens, floppy disks, or digitisers.

heads-up digitising Heads-up digitising is a data input technique. The user generally has a layer displayed on the computer screen and uses this data source to trace a new data layer. The equipment necessary includes a mouse, the computer and screen. This technique is often used to extract data from scanned images. For example, an aerial photograph may be displayed in digital format and the user may trace features (such as a coastline) into a new, independent layer. The phrase heads-up digitising is coined as the user is looking at the screen rather than bending over a digitising table.

hierarchical database A hierarchical database is a type of database (data organisation) that has a structure like a decision tree (or a family tree). Branches link shared data in a one-to-many relation.

imagery A GIS user or analyst would define imagery as a type of GIS data. Usually it refers to remotely sensed data, i.e. information about a surface gathered by an instrument that is not in physical contact with the surface. Photographic cameras and radar are well known imagery collection instruments.

imagery packages Imagery or remote sensing packages are software systems developed for the express purpose of processing, analysing, and displaying imagery data.

importing This is a process that is used to bring data into a GIS. Generally it involves digital data in a format other than the native format of the GIS software being used.

information   Information is an understanding extracted or learnt from data.

interface   Interfaces are connections between two elements. This may be a connection between the hardware and the software for the processing of commands, or a user interface between the software and the user for gathering user inputs and commands.

interpolation   This process uses known data values to predict values for unknown (unsampled) locations. Interpolation is often used with point data to create a surface of values.

keyboard   This is a hardware peripheral and an input device. It is a set of letter and number keys and an electronic sensor board, used in the same fashion as a typewriter.

kriging   This is a technique using for the interpolation of data across a surface.

Land Information System (LIS)   This is a system similar to a GIS, with the exception that the spatial and attribute data are related to land parcels.

lattice   A lattice is a DEM built with regularly spaced nodes. The nodes contain attribute elevation data. The lattice resembles a fish net draped over a landscape. The areas between nodes are voids, unlike the grid cell in a DEM.

layer   In this context the layer represents a basic level of data storage in GIS. A layer will contain one data type (point, line, polygon, or grid cells) and represent one feature (such as streams or dams). Each layer will have a unique set of spatial and attribute data.

line   A GIS line is a component of vector data structure. A line is referenced by a starting and end point (termed nodes) and intermediate inflection points (termed vertices). The connection of nodes and vertices creates a one-dimensional linear feature. Lines have a length attribute.

liveware   Liveware is a slang term given to the component of the system that represents the human user.

machine   Machine is a slang name for a computer.

mainframe   A mainframe is an interconnected system of users and one powerful machine which contains all the software and data.

map extent A map extent is the spatial extent of the data in a layer. This is usually expressed in terms of minimum and maximum x- and y-coordinates.

Map Grid of Australia (MGA) MGA is an Australian coordinate reference system used to locate data in space. The coordinates are derived from the Universal Transverse Mercator projection using an Australian Datum. An example datum point is the new GDA94, expected to be adopted in most Australian mapping by the year 2000.

merging (as a geo-processing tool) This geoprocessing tool creates a new layer, based on data in an existing layer, by removing unnecessary or superfluous polygon boundaries (generalising).

metadata Metadata is data about the data. Usually metadata includes items such as the date of capture into digital format, data ownership, and the scale of the original map product. Metadata is crucial in establishing the appropriateness of a data set to an analysis.

model A GIS model (or the process of modelling) combines spatial and aspatial data using the GIS toolbox, in an attempt to approximate reality. Models can be explicative, predictive, or statistical.

module A module is a section of a unified whole. A module of software is part of a larger program. A software module will have a specific function and may be able to operate in isolation from the other modules. Examples may include a mapping environment, a raster environment, or an analysis program.

mouse A mouse is an input device consisting of a small hand-held unit that has, usually, two or three buttons. The mouse links with a cursor on the screen. As you move the mouse along the desk, the cursor correspondingly moves across the screen. This enables the point and click method of selection and provides limited drawing capabilities.

neighbourhood A neighbourhood is an area surrounding a feature of interest. A neighbourhood may be defined within a raster grid or vector layer. Generally, the user determines the spatial extent of the neighbourhood based on knowledge of the application theory.

network A network is a set of interconnected linear features. In GIS, a network analysis involves investigating the flow of an entity (such as traffic or water) along these linear elements (roads or streams). These linear features may have a direction element and other attribute data relating to movement, such as barriers and impedance values. A computing network defines interconnected hardware and software data communication systems.

node A node is referred to in this text as an end point (start or finish) of a line. The term 'node' may be used elsewhere in GIS literature to refer to a label point in a polygon and even a point.

object-oriented database An object-oriented database is a form of database in which all elements are considered unique objects. Objects sharing similar properties and characteristics belong to the same class (instances of a class). The flexibility offered by this database type commends its adaptation to many GIS applications.

overlay Overlay is a process whereby two layers (or maps) with a similar or common map extent are merged into one product. The traditional method involves pressing two map sheets together and tracing features of interest by hand. The GIS process is computer driven and creates a new digital layer.

plotter A plotter is an item of hardware used to print out maps on paper. A plotter draws output as continuous lines, using pens or an ink spray.

point A point is a single location defined by a x-coordinate and a y-coordinate in a vector data structure. Points are the simplest unit in vector data structures and are used to build lines, which, in turn, are used to build polygons.

polygon A polygon is an enclosed area in a vector data structure. Polygons will have measures of perimeter length and area stored in the attribute table. Polygons are composed of connecting lines.

precision Precision refers to the level of detail in an item definition. In numeric terms, this may be a measure of the number of decimal places used in a spatial reference; e.g. 25.38756 is more precise than 25.4.

predictive modelling Predictive models attempt to use data and known interrelationships to predict a new state when one or more data elements are altered.

printer A printer is an output (hardware peripheral) device for producing paper product displays of maps or data. A printer is distinguished by the method it uses for printing. Examples are dot matrix, laser, and line printers.

projection Projection (verb) is an accepted mathematical process by which spatial data, which is part of the sphere of the earth, is approximated into a 2D data array. Any projection from a sphere into a flat 2D display will involve some distortion of reality. The manner in which the data are distorted depends largely on the type of projection used.

protocol A protocol is a set of rules dictating how data will be transmitted, usually across a network of computers.

proximity analysis Proximity analysis is a form of analysis in GIS that involves a query of the closeness of one entity to another entity based on a user-defined neighbourhood and a user-defined search distance.

raster data Data in a raster structure is cell-based data. The raster world is viewed as a surface of equally sized and spaced grid cells. Each cell contains an attribute that is (generally) representative of the feature occupying the majority of the cell. Examples of GIS data that are ideally treated as raster data are imagery and continuous data.

relational database A relational database is built on two-dimensional tables (relations). Relation tables can be linked or divorced from one another using any field (column) in the tables as a key. This is a popular form of database, although many other, newer types of database are gaining popularity.

remote sensing Remote sensing involves the gathering of information using a method that does not require the measuring device to be in contact with the entity being measured. Modern remote sensing involves sensors such as satellites and radar, and imagery such as infrared and thermal imagery.

resolution Resolution refers to the ability to identify features on a landscape. It is a measure of the closest distance between two unique, identifiable features. Resolution is frequently used to describe imagery data. A low resolution suggests a large raster cell size, and a high resolution suggests that the cells are smaller. A high-resolution form of imagery may allow the identification of features as small as a tree in a landscape, and a low resolution may make it difficult to identify a football field.

scale Scale relates the size of an entity on a map to its size in real life. It is usually expressed as a ratio of distance on the map to distance on the ground. 1:2000 indicates that 1 cm on the map represents 2000 cm on the ground.

scanning Scanning is a process undertaken by a device (scanner) that automatically captures information as digital data. A scanner operates in a manner similar to a photocopier, simply reproducing the original information. Instead of a paper copy of the original, the end result of scanning is a digital version.

screen unit A screen unit is a coordinate unit based on the position of the data on the monitor, or computer screen. Screen units vary, depending largely on the graphic display size and the map extent chosen to view the data, and are therefore little use beyond any one particular display. The screen unit, like the digitiser unit, is an artificial referencing system for data.

software Software is a computer program that stores the instructions telling the computer what to do when a command is issued or a button is pressed. Examples of software include GIS packages and database management systems.

spatial data Spatial data are data that can be said to vary across space. Space is usually expressed as a function of position along one axis (x-axis) and up/down another axis (y-axis). Microscopic particles distributed across a viewing slide have a spatial distribution. The spatial data would be encapsulated by their relative positions on the slide.

statistical modelling Statistical models are based on statistical knowledge or facts, an example being a probability model.

tool GIS tools are all the functionalities available within the GIS.

topographic map Topographic maps display the lie of the land and features on the surface of the land. Usually they portray a mixture of physical and social environmental features (layers), such as streams, contours, roads, and buildings of special interest.

topology/topological Topology refers to knowledge of the surrounding data elements or an understanding of the context of spatial and attribute data. Topological elements can be thought of as relationship or connectivity elements. For example, knowing that the library is situated between the post office and the supermarket is a topological statement. Explicit topology relates to vector data.

Transmission Control Protocol over Internet Protocol (TCP/IP) TCP/IP is a standard for controlling transfer of data between two different networks.

triangulated irregular network (TIN) A TIN is a digital elevation model 2.5D landscape, displayed using a triangular element. Triangles of varying size and area are used to represent planes across the surface of the land.

updating (as a geo-processing tool) This geoprocessing tool creates a new layer based on data in an existing layer by updating, or altering, a portion of the data (spatial or attribute) based on another, independent layer.

user-friendly A software program is said to be user-friendly if a non-expert can use it with ease. This implies adequate help documentation and an intuitive interface between the software and the user.

vector data Vector data are data structures for the storage of spatial data. Vector data are stored, illustrated, and analysed as polygons, lines, or points.

vectorising Vectorising is the conversion of non-vector data into vector formats. Commonly this involves changing raster data into vector data. An example may involve a scanned image defining locations of lakes. Scanning produces a raster image. The image needs to be vectorised into lines and then perhaps built into polygons.

vertex A vertex (vertices is the plural) is an inflection point in a line. A vertex is neither a starting nor finishing point of the line; rather, it represents an intermediate point where the line changes direction.

viewshed A viewshed defines the area visible from a particular point (or line, or polygon, or raster cell/s). The creation of the viewshed relies on digital elevation model data.

visual display unit (VDU) A VDU is the hardware device used for viewing the data and the GIS software interface. This is a screen that looks much like a television screen.

window A window is a rectangular area on a computer display screen that runs a program or displays data. The value of using windows is that multiple windows can be open at the same time. In terms of GIS this means that one window may be open to receive commands and another may be open as a display window for mapped data.

x-axis and y-axis The x-axis is the coordinate array that alters value in an east–west direction. The y-axis is the coordinate array that alters value in a north–south direction.

x-coordinate and y-coordinate These coordinates are used to define a spatial location. The location is referenced by a position in relation to the x-axis and y-axis.

References and further reading

Chapter 1

Burrough, P.A. (1986). *Principles of Geographical Information Systems for Land Resources Assessment*. Oxford: Oxford University Press.

Burrough, P.A. and McDonnell, R.A. (1998). *Principles of Geographical Information Systems*. Oxford: Oxford University Press.

Carter, J.R. (1989). On defining the geographic information system. In W.J. Ripple (Ed.), *Fundamentals of Geographic Information Systems: A Compendium*. Virginia: ASPRS/ACSM.

Chrisman, N. (1997). *Exploring Geographic Information Systems*. New York: John Wiley & Sons.

Smith, T.R., Menon, S., Starr, J.L., and Estes, J.E. (1987). Requirements and principles for the implementation and construction of large-scale geographic information systems. *International Journal of Geographical Information Systems* 1: 13–31.

Chapter 2

For an in-depth discussion of data, data types and data structures try Chapter Two of:

Burrough, P.A. and McDonnell, R.A. (1998). *Principles of Geographical Information Systems*. Oxford: Oxford University Press.

An easy reading discussion is also offered in Chapter Two of:

DeMers, M.N. (1997). *Fundamentals of Geographic Information Systems*. New York: John Wiley & Sons.

Chapter 3

Burrough and McDonnell give a succinct description of the process of converting existing documents into a digital format in Chapter Four of the text:

Burrough, P.A. and McDonnell, R.A. (1998). *Principles of Geographical Information Systems.* Oxford: Oxford University Press.

A paper written by Rhind, discussing copyright of GIS data, might also be of interest to the reader.

Rhind, D.W. (1992). Data access, charging, and copyright and their implications for geographical information systems. *International Journal of Geographical Information Systems* **6**: 13–30.

Chapter 4

Chapter Six of the text by DeMers offers a more in-depth, yet non-technical, discussion of errors and error types.

DeMers, M.N. (1997). *Fundamentals of Geographic Information Systems.* New York: John Wiley & Sons.

Chapter 5

Maling offers a very useful paper on coordinate systems and projections in GIS in:

Maling, D.H. (1991). Coordinate systems and map projections for GIS. In D.J. Maguire, M.F. Goodchild and D.W. Rhind (Eds), *Geographic Information Systems: Principles and Applications* (Volume 1). London: Longman Scientific and Technical.

Chapter 6

Readers with an interest in map design are directed to the paper by Buttenfield and Mackaness:

Buttenfield, B.P. and Mackaness, W.A. (1991). Visualization. In D.J. Maguire, M.F. Goodchild and D.W. Rhind (Eds), *Geographic Information Systems: Principles and Applications* (Volume 1). London: Longman Scientific and Technical.

DeMers provides a useful overview of data output in Chapter Fourteen of his text:

DeMers, M.N. (1997). *Fundamentals of Geographic Information Systems.* New York: John Wiley & Sons.

Chapter 7

The following references offer detailed explanations of database structures mentioned in this chapter:

Kleiner, A. and Brassel, K.E. (1986). Hierarchical grid structures for static geographic data bases. In M. Blakemore (Ed.), *Proc. Autocarto London.* Imperial College, London, Sept.

Lorie, R.A. and Meier, A. (1984). Using a relational DBMS for geographical databases. *Geo-Processing* **2**: 243–257.

Maguire, D.J., Worboys, M.F. and Hearnshaw, H.M. (1990). An introduction to object-oriented geographical information systems. *Mapping Awareness* **4**(2): 36–39.

More generalised explanations can be found in:

Frank, A.U. (1988). Requirements for a database management system for a GIS. *Photogrammetric Engineering and Remote Sensing* **54**(11): 1557–1564.

Healey, R.G. (1991). Database management systems. In D.J. Maguire, M.F. Goodchild and D.W. Rhind (Eds), *Geographic Information Systems: Principles and Applications* (Volume 1). London: Longman Scientific and Technical.

Burrough, P.A. and McDonnell, R.A. (1998). *Principles of Geographical Information Systems.* Oxford: Oxford University Press.

Chapter 8

There are a variety of texts available for a more detailed investigation of these elementary tools. One that has proven particularly useful to the novice is:

Davis, B. (1996). *Geographic Information Systems: A Visual Approach.* Santa Fe: OnWord Press.

Chapter 9

Once again, Davis provides an excellent follow-up to the material presented here.

Davis, B. (1996). *Geographic Information Systems: A Visual Approach.* Santa Fe: OnWord Press.

DeMers provides a more detailed explanation of many of the techniques introduced in Chapters Seven and Eight of his text:

DeMers, M.N. (1997). *Fundamentals of Geographic Information Systems.* New York: John Wiley & Sons.

An ideal reference for further reading on the use of overlay is the classic work of McHarg:

McHarg, I.L. (1969). *Design With Nature*. New York: Doubleday/ Natural History Press.

Chapter 10

A well-structured review of GIS overlay, with a section on raster overlay, can be found in:

Chrisman, N. (1997). *Exploring Geographic Information Systems*. New York: John Wiley & Sons.

Chapter Eight of the text by Burrough and McDonnell provides a more in-depth description of spatial analysis using continuous data.

Burrough, P.A. and McDonnell, R.A. (1998). *Principles of Geographical Information Systems*. Oxford: Oxford University Press.

Chapter 11

A follow on from the material presented here is provided towards the end of Chapter Seven of the text by:

Burrough, P.A. and McDonnell, R.A. (1998). *Principles of Geographical Information Systems*. Oxford: Oxford University Press.

An informative, though technical, reading regarding least cost path determination can be found in:

Douglas, D. (1994). Least cost path in GIS using an accumulated cost surface and slope lines. *Cartographica* **31**(3): 37–51.

Chapter 12

Weibel and Heller provide an informative and interesting overview of digital terrain analysis in:

Weibel, R. and Heller, M. (1991). Digital terrain modelling. In D.J. Maguire, M.F. Goodchild and D.W. Rhind (Eds), *Geographic Information Systems: Principles and Applications* (Volume 1). London: Longman Scientific and Technical.

A paper discussing the use of 3D (as opposed to 2.5D) GIS is provided by Raper and Kelk in:

Raper, J.F. and Kelk, B. (1991). Three-Dimensional GIS. In D.J. Maguire, M.F. Goodchild and D.W. Rhind (Eds), *Geographic Information Systems: Principles and Applications* (Volume 1). London: Longman Scientific and Technical.

The following also offer further information on the principles introduced in this chapter:

Burrough, P.A. and McDonnell, R.A. (1998). *Principles of Geographical Information Systems*. Oxford: Oxford University Press.

Beven, K.J. and Moore, I.D. (Eds) (1994). *Terrain Analysis and Distributed Modelling in Hydrology. Advances in Hydrological Processes*. Chichester: Wiley.

Chapter 13

A discussion of modelling in GIS is provided in Chapter Thirteen of:

DeMers, M.N. (1997). *Fundamentals of Geographic Information Systems*. New York: John Wiley & Sons.

Further reading relevant to kriging can be found in:

Oliver, M.A. and Webster, R. (1990). Kriging: A method of interpolation for geographical information systems. *International Journal of Geographical Information Systems* **4**(3): 313–332.

Tomlin and Berry introduce the use of cartographic modelling in:

Tomlin, C.D. and Berry, J.K. (1979). A mathematical structure for cartographic modelling in environmental analysis. In *Proceedings of the 39th Symposium of the American Conference on Surveying and Mapping*.

Tomlin has also produced a text on this topic:

Tomlin, C.D. (1990) *Geographic Information Systems and Cartographic Modeling*. New Jersey: Prentice-Hall.

Index